Worksteads

A Headlands Press Book

Worksteads

LIVING & WORKING IN THE SAME PLACE

JEREMY JOAN HEWES

photographs by
David Seligman

Dolphin Books   Doubleday & Company, Inc.   Garden City, New York

Created and produced by
The Headlands Press, Inc.
243 Vallejo Street
San Francisco, California 94111

Library of Congress Catalog Card Number: 80–941
ISBN: 0–385–15995–1

Printed in the United States of America
10 9 8 7 6 5 4 3 2 1

This book is for my family,
and for the memory of an old friend.

Contents

How _____ 77

What if... _____ 155

Appendices _____ 159

Acknowledgments

The efforts and talents of many people have gone into this book. Some of these contributors have worked behind the scenes, and they deserve recognition for their help in putting a concept between covers. My special thanks go to Linda Gunnarson, for her ideas and energy and caring throughout this project; Georgia Oliva, for giving visual life to these pages, for knowing from the beginning what worksteads meant, and for constant support; David Seligman, for the collection of "looks" in this book and for being a great traveler; Andrew Fluegelman, for his enthusiasm and for making this project possible; and Barry Traub, for his steady guidance and welcome suggestions. Sally Shepard contributed her bountiful research, John Edwards and Nancy Feiner lent their expertise, Emilio Mercado provided excellent film processing and printing of photographs, and Howard Jacobsen and his staff at Community Type & Design supplied quality typesetting and preparation of mechanicals. All of these professionals share in the credit for this intricate collaboration known as a book.

Another group of unseen heroes offered suggestions of worksteaders, gave general encouragement, and helped smooth the process of gathering information. These thanks go to Beth Battle, Peter Beeman, Rita Cahill, Bernie Choden, Pam Cokeley, Ellen Davis, Aileen Egan of Thomar Publications, Carol Hacker, David Fischer, Ann Flanagan, Laura Hewes, Trish Hibben, Ron Jones, Betty Kahn, Nancy Kales, Richard Mayer and Margaret Raymond of Artists Equity, Alan and Judy Miller, Laura Novick, Sarah Satterlee, Tom Sexton, Virginia Sloan, Dale Smith, and Diane Zimmerman Wells. The assistance and patience of my colleagues at Island Press also are greatly appreciated.

Finally, my thanks are gratefully extended to the worksteaders, their families, and their co-workers, who so readily shared their time and experience and thereby gave shape and substance to a new perspective on contemporary life. This book is the result of their generosity, and a tribute to them all.

J. J. H.

About This Book

The word *workstead* slipped out of my ballpoint as I was writing a letter to a friend who moved to the country and took his work with him. This word—and the idea—grew from the notion of a homestead, a concept that has always attracted me. The sense of independence, self-reliance, and pioneering that I associate with homesteads also quickly surrounded the new word. And with this meaning came a particular relevance to my life, my friend's, and many others—we aren't plowing the upper forty or cutting timber for a cabin, but we are carving out a territory with and for our lives. For me, this personal pioneering is far more rewarding than the traditional compensations of a salaried job, with its benefits, regular hours, and yearly vacation.

Worksteads is a documentary book. The information about home-based careers in these pages has come largely from the people who earn their livings at home, and we've tried to reflect their wisdom, concerns, and enthusiasm in our text and photographs. Finding worksteaders throughout the country was an informal process: we started at home, in San Francisco, and talked to friends and acquaintances who combine living and working; these people shared their experience with us and suggested other worksteaders. This network led us to all parts of the nation, as virtually every worksteader we met told us about one or two more.

The process of locating worksteaders, corresponding with them, and talking by phone resulted in our three-week trip across the country. With the help of a resourceful travel agent, more airlines than I knew existed, and uncommonly good weather, David and I canvassed the East Coast, from Maine to Washington, D.C., moved on to the Midwest, and visited the Plains states before heading back to the West. We were made welcome at every stop, and we spent a day or a few hours with each home-based businessperson. Our only regret was not being able to stay longer with each worksteader; our only frustration was seeing so much beautiful countryside and not having time to stop.

In all, we visited a hundred worksteads. Though the variety of vocations, settings, and incomes is great, there is significant common ground among these worksteaders. All have chosen the independent course of a home-based career, and all are enjoying life.

Of course worksteads aren't for everybody. Society still needs work places for concentrated activities, and many people choose to join a group work effort or purposely separate home life and work. My goal in exploring the options that a workstead offers is mainly to remind all of us that we have some realistic and exciting choices. We all can influence the larger course and the daily conduct of our lives by learning, planning, and choosing—and a workstead is an excellent choice for many of us.

What

Workstead is a new word, the natural joining of terms for "livelihood" and "surroundings"; it means living and working in the same place. Yet the workstead concept goes beyond the physical joining of home and work place, for it emphasizes a scale of activity that gives equal importance to a person's occupation and the essential people and comforts in his or her life.

Historically, worksteads were the rule. Sparsely settled lands, slow travel over primitive roads, and sluggish communications contributed to the logic of staying home to make a living. Although the Industrial Revolution marked a major shift away from the home as a work place, farms, some cottage industries, and home offices of doctors, dentists, and other professionals remained traditional workstead situations.

Today, worksteads include more than traditional occupations. Technology has made possible sophisticated production in diverse locations. The roads are good in most places; the telephones and computers are better. So a product or service can originate in a woodland studio or a studio apartment and still find its way across the country or the oceans.

Worksteads are established for many reasons and from many circumstances. For some people, a workstead is a necessity. Particularly in isolated areas or in rock-bottom-income vocations, a person must live among his tools or set up his easel in the alcove with the best light. For others, working and living in the same place just feels right. The worksteader simply doesn't want to spend an hour or more traveling to a job or catching only a few hours each day with family and friends.

Home work places are especially valuable in this era of dwindling resources and rising costs. The savings of such precious commodities as time, fuel, and money are significant for every workstead family, and as the number of worksteads grows, these savings will have an important impact on society as well.

That worksteads are feasible and timely seems clear; that they are also enormously beneficial to their participants is the payoff. There is a personal dimension to worksteading that surpasses convenience or necessity. When a person lives where he works, he is more able than the 9-to-5 worker to participate fully in family activities, crises, and growth.

Perhaps most important, the worksteader usually doesn't make a distinction between "living" and "working." She is more fully "herself" and need not don a uniform or conform to a regimen that often accompanies a job away from home. This is the primary appeal of a workstead to many people—it enables them to live in ways that are satisfying to them and to their companions. And people who are satisfied with life have a richness that will rub off.

Why

Each of us has felt the pinch of a changing economy and the unsettling shifts of a world in transition. In response to these changes, or in anticipation of greater ones, many people are fashioning new lifestyles or returning to long-neglected forms of livelihood. A workstead is one alternative that forges the old and the new by combining living and working in the same place. The workstead movement has begun spontaneously—in many places and among many different people, without leaders or sponsors or media blitzes. As many as five million individuals have decided to combine living and working, each for his or her own reasons and in his or her own style.

Some common concerns are at the heart of this vibrant and growing movement. Economic factors are certainly a primary impetus for worksteads, as well as the most tangible measure of the movement's impact on society. In addition to its money-saving features, a home-based career offers the individual comfort, freedom from commuting, a sense of independence, and the opportunity to be closer to family. Collectively, worksteads offer society the example of entrepreneurs who emphasize conservation of resources, question the use of energy-intensive technology, and urge a scaling down of most institutions.

In short, worksteads are the ground where a certain kind of economics meets a certain kind of humanism. The people who combine living and working have acted to improve the quality of their lives, and their movement is flourishing, so a cogent look at the reasons for choosing this alternative should be instructive to all of us in shaping the future.

Money and Time

Money is a principal consideration in most worksteads. A home-based career offers the opportunity to add to the family income without extra spending for space, or to start an independent business at a greatly reduced financial risk. The federal Small Business Administration, for example, advises people who are beginning new enterprises to work from home if possible—to keep their overhead low and to stretch their capital farther than if they were paying rent on business property.

There are trade-offs, of course, in using a work place that accommodates living quarters or a home that doubles as an office. Private living areas may be sacrificed to the business and office space may be cramped, but the economic advantages of a workstead business almost always make these compromises worthwhile. In many instances, the money saved literally makes the venture possible, whether someone is quitting a 9-to-5 job to work solo, trying out an experimental product or process, or beginning a new career.

Worksteads offer other savings, too. Anyone who commutes to a job is painfully aware of the rising costs of gasoline and public transit, as well as the time and exasperation involved in getting to work. Census Bureau figures show that on a typical work day, Chicago-area commuters spend a total of 1,056,133 hours traveling to and from their jobs. In Atlanta, one day's commute totals a mere 219,000 hours—still an amazing waste of good time.

Resources

Fuel savings and reduced pollution are the larger benefits of a reduction in commuting. In his lively view of the future, *The Third Wave,* Alvin Toffler cites a study estimating that if 12 to 14 percent of the commuting population had stayed home to work in 1975, the United States would have saved enough fuel to offset gasoline imports for that year. And since automobiles are the source of 60 to 85 percent of all air pollution in this country (with the highest concentrations in urban areas), reduced commuting would mean easier breathing for everyone and less danger from pollution "fallout." Fewer people traveling also means fewer deaths and injuries on the roads and lower road maintenance and construction costs.

Worksteads help to conserve other valuable public resources by extending their useful lives. Buildings and dwellings that have outlived their original purposes have been converted to working and living spaces, which is especially important in this era of burgeoning land and construction costs. Moreover, twenty-four-hour occupation of these existing structures constitutes a truly efficient use of space, utilities, and public services such as fire and police protection.

"My primary aim in work is to be able to control my own time. I worked at several jobs, one as an editor in New York for a year, but I felt imprisoned in the confines of those offices, with other people determining how I was going to spend my time instead of my determining that. It was really uncomfortable for me.

"Now I can work any time I want to—and that's not always at the regular times. In the middle of the day, I can go out and take two hours to just enjoy the world. That's the thing I like about it—that freedom." —*Candice Jacobson, publicist*

"If a person is going to leave a job to work at home, he needs a very clear attitude about how he's going to live. I set up a rather modest goal of the kind of security I wanted to have before I left the law firm. I don't buy expensive clothes, for instance. I enjoy cooking, so I don't go to restaurants much. If you have a place to live, where you can also work, you can get along on very little. The rest of life doesn't really take too much money, if you have a place to be." —*George Hellyer, attorney*

By conserving resources and maximizing the use of space, worksteads are harbingers of the vast changes that many observers have predicted for the planet. Certainly this concentration of work and living in one place exemplifies the "small is beautiful" ethic of economist E. F. Schumacher. Home-based businesses also contribute to a decentralization of cities and a scaling down of industries, processes that social theorists and environmentalists see as inevitable. Industrialism and urban sprawl have damaged the environment so thoroughly, they argue, that this unabated growth must be reversed, with some observers predicting that at least half our population and industries will be located in rural areas by the next century.

Technology

The anticipated rural resurgence will come about largely as a result of advances in technology. Sophisticated telecommunications devices and networks that utilize computers, telephone lines, and satellites will make the workstead alternative feasible in nearly all locations. Already, the miniaturization of electronic components has resulted in a variety of products that fit easily into a home office and operate on household electricity. For example, a chip that is smaller than a fingernail can store up to a million bits of information in a computer the size of a pot roast.

Worksteaders are just beginning to take advantage of this telecommunications boom. In late 1980, about two hundred thousand home computers were in use, many of them already plugged into data banks that include the entire United Press International news service, airline schedules, and a 10,000-item catalog shopping service that takes computer orders. On the horizon are worldwide data banks, electronic filing systems, and a voice-operated typewriter that will make transcription simultaneous with dictation. All of these innovations will encourage independent workstead businesses and facilitate more work-at-home programs for employees of large organizations, some of which already have established such programs on a limited basis.

Worksteads have played a significant role in the creation of advanced home technology, as well. One of the first home computers was developed by Apple Computer, Inc., a firm founded in 1976 by two engineers who wanted to buy a "hobbyist" computer and couldn't find one on the market. The two men built the original Apple in the spare bedroom of one of their homes—raising their initial capital by selling a calculator and a van—then moved to a garage when operations outgrew that bedroom. Now Apple is a leader in a home computer market that will do $5 billion in business by 1984. Of course the firm no longer operates out of a garage; its plant and office employ four hundred people and cover 80,000 square feet of space in Cupertino, California—the heart of Silicon Valley.

"In terms of the future, it seems to make sense to go toward multi-use and multi-functional neighborhoods as a planning direction. The whole planning theory in cities has been based on the car, and now in this energy decline we have to look at that again. The ideal vehicle for the future is a three-wheeled rickshaw—with pedals and an electric assist that works on batteries.

"Industry is a misleading word—because industry suggests smokestacks or assembly lines—and there are a whole bunch of soft industries that are information-oriented or technologies that have no pollution whatsoever. Or electronics assembly, which is a quiet industry that can be done on a small scale. These kinds of industries could be right in our neighborhoods, right where we live. That's what I'm trying to demonstrate in this old warehouse. I've always lived where I work." *—Peter Ziegler, Earth Lab Institute*

"When I'm working on the computer, basically I'm learning—I'm teaching myself. In other jobs, I've worked mostly on the logical level and worked with people, trying to design how they would use a computer. Now I'm solving problems directly with this piece of hardware that I put together.

"Having my own computer has changed my home life to a certain extent, too. I now spend more time at the computer instead of watching television. I consider that very positive." *—Buck Kales, computer consultant*

Many worksteads are also fertile ground for developing "appropriate technology," which makes use of non-resource-depleting forms of energy such as wind and solar power, as well as methane and hydrogen fuels. In addition to providing or augmenting the energy supplies for both rural and urban worksteads, these small-scale technologies are the focus of several work-live communities presently conducting research in self-sufficient living, including the Farallones Institute in Occidental and Berkeley, California, and the New Alchemy Institute in Falmouth, Massachusetts.

Self-Reliance

A new breed of generalists is emerging, people who want "hands-on" experience in all facets of their lives. The most obvious example of this renewed impulse toward self-reliance can be seen in the family farms and modest homesteads that have been established in the last two decades. These new farmers—many of whom have come from urban jobs or professions—are carrying on the agricultural tradition that was America's foundation for nearly two centuries. Their wish for simplicity and the rewards of hard work on their own soil often has been a response to the intense pace and competition of the conventional working world.

People who've chosen to attempt self-sufficiency in a rural setting often are accused of "dropping out." But writer Wendell Berry, himself a farmer who uses horse-drawn tools, suggests that he is "dropping *in* to a concern for the health of the earth." In fact, Berry's analysis of his situation aptly summarizes what many worksteads mean to their participants: "One's home becomes an occupation, a center of interest, not just a place to stay when there is no other place to go; work becomes pleasure; the most menial task is dignified by its relation to a plan and a desire; one is less dependent on artificial pleasures, less eager to participate in the sterile nervous excitement of movement for its own sake; the elemental realities of seasons and weather affect one directly and become a source of interest in themselves; the relation of one's life to the life of the world is no longer taken for granted or ignored, but becomes an immediate and complex concern. In other words, one begins to stay at home for the same reasons that most people go away."

Satisfaction with Work

An ailment common to many job-holders—a lack of satisfaction with their work—is also a motivation for home-based careers. In his book *Working,* Studs Terkel points out that too often the good worker finishes last: "A farm equipment worker in Moline com-

"One thing about farming is that your rewards are immediate and very obvious. They're visual—you do all this work, and you go out in the vineyard, and you really do see the fruits of your labor every single year." —*Chip Lyeth, vineyard owner/ farmer*

"It's a way of life that's very meaningful. We don't like to buy vegetables—we eat seasonally. We eat what's ready in our garden, which means that we may have broccoli for two months and not have a tomato until summer. But it's so satisfying—it makes sense to use what we have and what we can do ourselves." —*Cathy Lyeth, vineyard owner/farmer*

"This isn't a very up-to-date way of farming. We're supposed to have push-button machines to do all this work. But we're old-fashioned—we just never went to all the new things. My dad always said never get in debt, so we just do pretty much what we can afford.

"When I was sixteen, I got out of high school and started to walk behind a harrow. I'm fifty now, so I've been at this for thirty-four years. You've got to have a love for the land, or you wouldn't stay here and take all these risks. You've got to fight all the elements. When I die, I'd like to go somewhere where there's no mud." —*Dick Molony, farmer*

"I would probably set up a studio before I would set up my bed, because it's a bigger need. It just never dawned on me that you don't live in the place you work. Working for somebody else is like turning the volume down. There's a certain amount of image that the boss wants you to exhibit in terms of being busy, and I'm not involved in that sort of charade. It's confusing enough to figure out some way to honestly represent who you are. Your family and the people around you are the ones who sustain you—not some job." —*Donna Billick, painter/ceramic sculptor*

plains that the careless worker who turns out more that is bad is better regarded than the careful craftsman who turns out less that is good. The first is an ally of the Gross National Product. The other is a threat to it, a kook—and the sooner he is penalized the better." This out-of-kilter reward and penalty system has influenced the revival of cottage industries, which are especially visible among craftspeople who take pride in making quality products by hand.

Comfort and Convenience

There are plenty of worksteaders who simply want the living to be easy—to roll out of bed, climb into yesterday's jeans and sweater, and stroll a few steps to the office. For many people, home is a magnet whose attraction is stronger than the lure of a potential high salary or powerful position elsewhere; these people often forego such rewards in favor of the comfort and convenience of a workstead career. Yet another aspect of a home career is mentioned by almost every worksteader—the chance to dress casually, which reinforces the sense of ease and comfort of being at home.

A sense of place that is scaled to personal dimensions is another attraction of a work-live combination. In a tongue-in-cheek look at urban life, journalist Charles Kuralt describes the consequences of a place that is out of scale with its residents. He bemoans the plight of a Manhattan commuter who wakes up in Greenwich Village, gets in a taxi or subway, and travels through seven zip codes on his way to a midtown office. Kuralt labels the commuter's condition "zip-code lag" and prescribes a visit to Spring Green, Wisconsin, where the same zip code (53588) goes on for miles.

Family

Worksteads can make the family a valuable economic entity. Sociologist Theodore Roszak observes that we may become increasingly dependent on the home economy as the urban-industrial economic base weakens, so that entire households will be producing goods to sell or trade for their livelihood. He advocates this return to home businesses as a means of preserving family structure: "I submit that any family-saving strategy which deserves to be taken seriously must include some provision for bringing work back into the home and for increasing, at least in some small measure, the economic autonomy of the family. Even our marriage counseling would do well to consider the importance of practical shared labor (by which I do not mean hobbies, games, or camping out) as an integrating force in the life of the family."

A workstead provides this "integrating force," as well as the

"By working at home, you bring in interesting people for your kids to meet. I think a very important part of a kid's education is to meet adults and see how they look and how they act—it's not just seeing relatives or friends' mothers and so forth, in the same domestic roles all the time.

"And I like the fact that the kids are around a lot. They're aware of who their daddy is and what their daddy does—I don't go and do this mysterious thing; I don't go away and come back. And I'm nourished by their presence. I like that sense of warmth and love suffusing what you do—it's a good thing to have in your writing and your mind.

"One day somebody called and was asking me to do a book, and while I'm trying to talk, my little one is pulling at my pants because she wants me to play house, and she wants to know whether I'm going to be the baby or whether she's going to be the baby and I'm going to be the mummy. There was some sense that what was happening in the ear and what was happening down around the knees were definitely two different worlds that ordinarily would never mix." —*Malcolm Margolin, writer/publisher*

"Being in a home is a much more relaxing situation. I used to work in a rehabilitation clinic, where there was a waiting room and a million kids, and the doctors would be walking around with white coats, and there would be screaming and fighting. By the time you got the child, it took a half hour of relaxation and deep breathing to get them to the point where it was conducive to do therapy. Here it's quiet, you can schedule them one at a time, and you can talk to the parents in private.

"I've gotten a lot of very positive feedback about how comfortable it is to be in a home. The kids are very relaxed and truly enjoy coming here—it's hard to get them to leave." —*Donna Scheib, speech-language pathologist*

opportunity for a family to be together for more than just evenings and weekends. The working members of the family still must do their work, of course, and the nonworking members must respect those efforts, but everyone who is part of a workstead shares a sense of unity and a feeling of accomplishment that are seldom possible for work-away businesspeople and their stay-at-home families.

Many worksteads also create "expanded" families—groups of friends and colleagues who are not related but who establish ties that are strong and lasting. In her book *Families,* Jane Howard argues for this broadened definition of family, since only 17 percent of American families are the traditional Mom-Dad-and-kids "nuclear" variety. In this age of separation and mobility, Howard asserts, we must make families of our friends, because the need for this association is always present.

Community

The impulse to belong to a larger group also motivates many people who establish worksteads. A home-centered career often provides three circles of association: the company and participation of children and mates in work; an expanded family comprised of co-workers and friends; and participation with neighbors in community projects.

The community link is especially rewarding to worksteaders who formerly left home to work at jobs elsewhere. The time saved by not commuting and the flexible work hours of home-based businesses allow them to attend more school functions with their children, take part in local government and business organizations, and develop a network of neighborhood contacts that can be mobilized for political efforts, social activities, or personal emergencies. These neighborhood networks benefit the larger community, too, by reducing alienation and promoting joint efforts for the common good.

Pioneering

For both individuals and society, worksteads represent a bond between contemporary people and the roots of the American experience. Worksteads are one way of pioneering in a modern culture that prizes large-scale industries, material consumption, and pursuit of wealth, often at great cost to the natural environment and individual well-being. The value of latter-day pioneers who are striving to get by with less of everything, including money, is emphasized by author Warren Johnson in *Muddling Toward Frugality:* "They are doing a task that is essential for our future, developing new skills and ways of living that will provide models for others as necessity

"Having the studio at home has been a merging of my vocation with my avocation. Being a history buff, I appreciate the challenges of the pioneers, and I think we're just pioneers in a different way. You know how we're raised in this educational system—somehow we're conditioned to think that there's a how-to manual, that somewhere there's a book that'll tell you how to do everything. It's inhibiting, because people won't do things because they think they don't know how—when nobody knows how to do them. New Dimensions is an example of something being created out of nothing, out of an idea."
—*Michael Toms, New Dimensions Radio*

"There's a real benefit for me in living and working here. I tend to have a basic shyness about meeting new people. But living here has forced me to develop that part of my personality that tends to want to hang back and not go ahead. I can't do it here—there's no way to hang back. That's been a real growth process for me."
—*Justine Toms, New Dimensions Radio*

"It's not a chrome and Naugahyde place—this office has the aura of a home. It's not intimidating. I think that's made a difference in the kind of practice we've developed, the kind of practice I want. And, because everything I need to do in a day is right here, if I have to wait for a patient, I can go in the house or go work in our garden."   —*Les Michaels, M.D.*

"Patients come in who also like to garden, and Les will take them out back and have them garden with him. He gets us involved in other ways, too. One night I came home from shopping and there was a baby basket in our family room. A patient had called to ask if we knew anybody who could babysit for her, and Les said, 'Well, Peggy's out, but why don't you bring the baby by?' We always have a crib or cradle here, and Les is easy about that sort of thing—people are amazed."   —*Peggy Michaels, birth counselor*

pushes more of us in that direction. Nothing could be more important. The pioneers are opening up new economic territory where subsequent settlers can join them."

The Inner Stance of Hospitality

Worksteads foster human connections that can't be measured by sales figures or profit margins. They counteract our modern tendency to build boxes around ourselves—houses, office buildings, automobiles—by investing business dealings with the warmth and personal touches of a home. This built-in hospitality enriches everyone associated with worksteads and carries forward a tradition reflected in American Indian cultures, according to Sister Maria José Hobday, a teacher who is a Native American. She suggests that cultivating "the inner stance of hospitality" is the worthy work of a lifetime: "If you and I are fundamentally not scared of living, if we've been through a few experiences of being thrown out on the rim and have survived, then we are willing to take that risk of welcoming, of inviting ideas and people and their ways and their experiences into our lives and our thoughts."

Worksteads help to provide that common ground for sharing lives and thoughts in an increasingly complex world. They offer essential lessons to our changing society—as independent businesses, as people worthy of notice, and as superb examples of the initiative and options that thrive in America.

"This whole business is very personal. That's why it's important to have my home reflect me to the people I work with. Instead of hauling all the food and equipment someplace, people come here and we can discuss it and I can have everything set up for them. And it gives me credibility, too, that they see me, that I live here, and that I have a large enough kitchen to work in. It gets into a basic feeling that I'm serious about what I do.

"It's a very personal thing—how you work with people. A lot of people can do the same work, but it's how you relate to people that counts. For me, it's not just cold business; there's a warmth to it, a certain rapport." —*Sandra Learned, food stylist/caterer*

"The way publishing works, your authors often get to be friends as well as business clients. They're coming to your home instead of to your office, so there's a little different feeling. An author who's a friend will come to discuss a chapter, and the next thing we know, it's eight o'clock and we're all in there having a great time talking about politics, religion, whatever.

"I also like being where my family is. Anita and the children will see me in my office at night—I may be reading manuscripts, or just sitting there dead tired. What I've said to the children is, 'After five o'clock, don't think of it as an office, think of it as our library.' I often ask them, 'Will you guys come in here and keep me company?' But that hasn't worked—the dog keeps me company." —*Ernest Scott, San Francisco Book Company*

"The girls never hesitate to come in and ask questions or have a conversation—not just at night, but anytime. And Ernest and I are always going into the kitchen for coffee or running into the other rooms for something, so there's a dialogue going on all the time with the children." —*Anita Scott, San Francisco Book Company*

Who

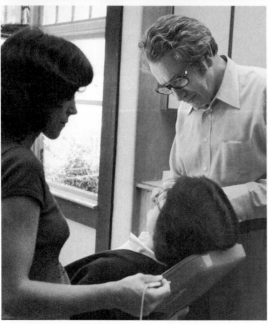

Worksteaders are some of everybody—people of all ages, living in many places, doing most kinds of work. Some workstead occupations have been with us for generations, such as farmers and ranchers, or the doctors and dentists who see patients in home offices. And there must be thousands of playrooms and basements that are used as music or dance studios every afternoon by teachers who give lessons to aspiring young performing artists.

These traditional home-based jobs still flourish, but the scope of worksteads has grown tremendously in recent years. A commercial recording studio, for example, is an elegant home for its owners; a nineteenth-century house is the site of a physicist's laboratory; and a one-time cabinet shop in the side yard of an Indiana home has grown into a furniture factory covering an acre. Worksteads are not limited to independent businesspeople, however. Large companies, especially in the word-processing field, have found their employees responding eagerly to experimental home-work programs, which are made possible, in part, by the growing use of computers and other sophisticated communications equipment.

Each of the worksteads presented in the following pages illustrates an individual adaptation of skills and resources to location and needs; together they reflect the diversity of the growing workstead movement. The worksteaders profiled range in age from twenty-eight to eighty-three and in education from elementary school graduates to Ph.D.'s. Some have set independent courses for their careers, such as the physicist; some have changed careers, such as the "magazine doctor"; and some have turned a dream into a livelihood, such as the couple who restored an old hospital as a guest house and art workshop.

Yet for all the variety of their jobs, settings, ages, and circumstances, these worksteaders have certain characteristics in common. All are practicing believers in the worth of small enterprise and the value of personal effort. They are people who have chosen to make work a close companion to living, and they are greeting the future on their own terms.

Martha Stuart
VIDEO PRODUCER

"I think your work should be really enjoyable. Some people don't use their work; they put their time in, and their home is where they have their pleasure. The idea that one is pleasurable and the other isn't, is ridiculous. It's not separable to me—everything should be pleasurable."

Martha Stuart lives and works in the top two floors of a Greenwich Village townhouse, where she and two assistants handle most details of the production and distribution of a series of video programs called *Are You Listening?* Martha has worked independently as producer and originator of this open-ended series since 1963; her programs have been shown on the national PBS network, on many local television stations, and in libraries and schools throughout the world. "I work in a strange

sort of cottage industry," Stuart says. "The best machines are rented by the minute, and you wouldn't want to own them—they're too big and complicated. So I do that work in a studio, and my day-to-day work—money-raising, distribution, things like that—out of my home."

A single parent whose two children are now grown, Martha began her workstead business when her son and daughter were just three and six. "When I set this business up," the producer says, "the most important thing was feeling that I owned my own time and that I could connect with the world, but at the same time be the sort of mother I wanted to be. So other things had to be arranged around that, instead of the other way around."

In fact, Stuart's children have become involved in her firm, Martha Stuart Communications. "Having my business here made things better between my children and me, and they share in the work and the pleasure," she notes. "When they were younger, I gave them 10 percent of anything they sold to a customer when they answered the phone, and so they've participated and made money all along. And then I thought, 'Well, the business is in our house—how can I fix it so it's not something that they feel resentful about because they can't just horse around in the house?' And so I gave them each 20 percent of the business, and they still own that. So it's all of our business—it's literally our cottage."

Martha's home serves as the center

for a variety of administrative and distribution functions associated with her video programs. One former bedroom has become a library, where clients often come to view and select programs. Another room is the publicity and telephone center, where associate producer Sherry Delamarter has her desk. Stuart holds meetings in any space that's unoccupied—often in her own bedroom or at the dining room table. All of these rooms are alive with color —paintings by friends and the folk art of the many countries Martha has visited for *Are You Listening?*

Frequently, too, Stuart will prepare a meal for her business associates, because she particularly enjoys mixing her pleasure in cooking with her pleasure in work. "The people that I know in my work are also my friends and kind of like my family," she says. "We take time out from work if it's someone's birthday or a holiday—I just think, it's my life, it should be fun."

Martha also finds that having people see her at home is good for new business relationships. "It helps people to know who I am," she points out. "I'm already strange enough to them; I seem very tough, and I'm divorced, and I'm having a lot of fun, and I live in Greenwich Village. There are twenty-five different reasons why people can wonder, 'Well, is she really businesslike, or is she this or that?' But when they come here and get a glimpse of my lifestyle, I'm sure that people move beyond a lot of the suspicion of me."

As her business has grown in the last few years, though, Stuart's need for private work time occasionally must compete with her open-door policy. "There are times when I feel invaded," she acknowledges. "If people are here for a screening and I'm busy with other work, they still like to say hello, and that can be a drain on my time. I don't like to be rude, but sometimes I don't feel as private as I'd like." She hopes to remedy this situation by renting the first floor of her building, which would become the front office and screening room for the business.

"We use ourselves as communicators," Stuart says, summarizing the philosophy behind both her work and her workstead. In her industry, especially, Martha believes that workstead businesses are essential to the future. "I think more and more that this is what people are going to want to do. Already, I work with many people who also work from home. One person driving to work in a car in the morning is crazy."

Peggy Swan GUEST HOUSE & Howard Levine ART WORKSHOP

"It was our dream to have this. The house itself is just supposed to survive—it isn't meant to make a lot of money. If we can continue to live in this style and with this much fun, we'll keep it going and not change things. We tell people they're contributing to their own enjoyment."

The Swan-Levine House is a kind of participatory guest house and art workshop, Howard Levine says. The people who visit this hundred-year-old Victorian mansion in Grass Valley, California, are encouraged to join in preparing meals, decorating for a friend's birthday party, or

working in the carriage-house art studio. Levine and his wife, Peggy Swan, are printmakers who had long wanted to combine their three strong interests in one location—continuing their careers as artists and teachers of printmaking, participating in their three children's

education, and having a family-oriented home to share with friends and strangers.

In 1975 the couple found their three-story building in California's historic Gold Country, and after one visit they bought the rambling 8000-square-foot structure. Built in the 1870s as a family home, the building was converted into a private hospital in 1900 and operated as such until 1968. Peggy and Howard have completed a major renovation of all three floors, restoring eight of the upstairs rooms to their original Victorian character and filling the house with period furnishings. "Finding furniture is a never-ending search," Peggy notes. "We bought many pieces in antique shops in the years before we got this house, and we still go to garage sales whenever we can." An ornate rosewood piano that dresses up the living room is on loan from a friend, and a few old wooden wheelchairs are the hospital's legacy to this new incarnation.

Restoration of the twenty-room house also seems never-ending, Howard relates: "It's a question of whether you tear everything out and start over, or you wait for something to break and then fix it. It's hard to tell which is better." During the family's six years in the house, they've tried both approaches. The original work to make the top floors livable was tear-out-and-start-over; Howard and Peggy assembled a crew of six people to help them, and the basic job took nine months of full-time work. Their bargain with this crew was free room and board in exchange for labor, and the

system worked well, although the group "ate up" a significant share of the money Peggy and Howard had allowed for this project. The funds they'd set aside from Howard's share in a family business was intended to last three years—it barely covered the renovation and crew costs. But this intensive learning experience in carpentry, wiring, plumbing, and other construction skills now permits Levine to make minor repairs and to judge when more expert help is needed for the fix-when-broken approach.

The Swan-Levine House presently pays its own way; that is, the twenty-five to thirty guests per month are sufficient to cover the basic maintenance costs for the building. The balance of the family's income is supplied by Howard's wholesale restaurant-supply business, which is conducted from his home office. The sale of the couple's art also supplements their income; Peggy, who signs her art with her full name, Margaret Warner Swan, is especially active in printmaking and drawing, and guests buy much of her art from the walls of the house.

Peggy's and Howard's children also participate in the family's enterprises. "Mara, who's four, comes out to the studio with me a lot," says Peggy, "and Michael, who's seven, often comes, too. They play or draw, and they're both actually capable of helping me." Nine-year-old Nathaniel gives tours of the house to guests and tells them stories he remembers from the remodeling phase. "This atmosphere provides lots of adult company for the children," Peggy adds, "and they love playing host to our guests' children."

Both Peggy and Howard volunteer regularly at their children's school, which is located across the street from their home. As Peggy points out, "We are a family that likes to spend a lot of time together. When Howard was working away from here, his time left with me and the kids was really small. We're much happier with his being at home and doing things from here."

The couple is also active in community affairs; Peggy serves on the grand jury, and Howard is on the school board and the local arts council. The house itself is considered something of a community asset and has been used for meetings and special events.

Although their setting might suggest wealth, Peggy and Howard live modestly and find that the costs of two independent businesses create a

definite financial strain. "Because we're self-employed," Howard says, "we provide our own benefits. We have to pay for health insurance, and we have a commercial insurance policy for the guest house." Peggy adds that these are the things people tend to overlook: "When you make your decision to go into this, you don't think of all these little expenses that add up."

Yet both Peggy Swan and Howard Levine agree that they have established the life they want. "There's so much to learn about politics and community in doing this," Peggy says. "Our house is a real focus. It's a statement about life and art and how we feel about living. And however much money we had, we would do the same thing, on whatever scale we could afford. If we could get a castle someplace, we'd do that."

Tom Gibbs
SCULPTOR

"There are so many different things I can do here. My job is not very specialized compared to other people who work alone. I have the option of working on a big steel sculpture and beating the heck out of it with monster tools; if I want to take a risk, I can climb up on top of it and weld. If I get tired, I can go to the other studio, put on my little coat, and sit and make models—like a gentleman artist."

Eight years ago, sculptor Tom Gibbs bought a three-acre hilltop in the middle of Dubuque, Iowa. The site had an abandoned stone quarry, the crumbling remnants of a stone barn, and a house so neglected that it was knee-deep in trash and weeds. The road to the house was so washed out that the realtor refused to drive his car up it. Somehow, Gibbs knew this was the home for him: "Part of my job is to be able to see the core and not to be influenced by the superficial things. So when I saw this place, I wasn't a bit depressed." What the sculptor saw beneath the trash and decay was an unparalleled site for his large studio, the makings of a comfortable house for his family, and a peaceful setting in which to live and work.

Tom and Dorothy Gibbs decided to buy the property "within ten minutes," he says, and immediately began to repair the house. Their daughter, Jennifer, stayed with relatives for the first few weeks of remodeling, because the living conditions were so spartan. Within two months the couple had made the home livable, and Tom started planning his big studio.

Gibbs had to fight the city bureaucracy—principally one ill-tempered inspector—to get permission for the large building, because although a Dubuque law recognizes the right of a citizen to work at home, it specifies that his office must be attached to his house. Tom, however, had seen immediately that the quarry area—not the house—was the perfect spot for a studio that would have to support steel sculptures and building

equipment weighing up to twenty tons. "The abandoned quarry is bedrock," he points out. "That means it's absolutely solid, and it means I didn't have to pour at least a foot of cement for a foundation—I could put all my heavy machinery there safely, and it wouldn't disturb anybody." Ultimately Tom did get the city's permission for the studio, chiefly because he challenged the law publicly and received local media coverage, as

well as advice from a New York arts group.

Like his large sculptures, the studio Tom constructed to house them is monumental. He borrowed scaffolding from a nearby quarry and hired two high school boys to help him put up the cement-block walls of the studio, which is twenty-five feet high and covers 1500 square feet. Then Tom began work on the overhead crane that could lift and move his machinery and his sculptures, which weigh as much as fourteen tons. "I put all that iron up alone, which I'm kind of proud of, because ironworkers on a regular construction job won't do that alone —it usually takes three guys," Gibbs reports. He built the superstructure to hold the overhead crane, which moves the length of the building on railroad tracks—all of this twenty-five feet in the air.

Then Gibbs moved in the heavy equipment with which he fashions the sculptures that are displayed in urban plazas and on university campuses. His studio houses a metal lathe, two power hammers, a small forge, a variety of clamps, tools, and anvils, a welding setup, and a huge wood-burning furnace. He stores most of his materials outside, and behind the studio building are two cranes and several completed sculptures.

More recently, Gibbs built a second, smaller studio on the site of the old stone barn, midway between the quarry and the house. He was able to use many of the stones—originally from his quarry—to reconstruct the back wall of this building, and the

structure is partially cut into the hill so that it will retain heat in winter and stay cool in summer. The sculptor uses this studio for drawing and making wax models for his pieces.

In planning both of these workstead structures, and in his approach to living in general, Tom has been very conscious of energy use and conservation. "I've made these buildings so that they can stand a cold shutdown," he says. "There's no plumbing in either one. The small studio won't freeze even if I don't heat it for weeks at a time, even in below-zero weather. It's partly underground; it's facing south, so the sun comes in and warms it; and these walls are thick—the back wall alone weighs eighty tons. The floor weighs fifteen tons. So when it gets warm, it's going to hold the heat."

The sculptor's building designs reflect both resource-consciousness and pragmatism, as he points out: "I'm not dedicated to energy conservation as a religion; it's more from a practical point of view. I don't have a lot of global impact—I look at it more in personal terms." When the area had a hard winter and a natural-gas shortage several years ago, Tom recalls, factories were closed and their inside temperatures were maintained at 45 degrees to keep the machinery from freezing. "Maintaining 45 degrees is a shutdown for them," Tom laughs. "That same year, 45 degrees was my operating temperature. In the daytime I turned the thermostat up to 45 from 25 or 30 degrees. This was before I had the wood furnace; I was heating with propane. The year before, I was operating at 35 degrees—so I work at a temperature at which factories

are shut down." Now Gibbs heats both studios with wood, burning scraps and parts of trees that are unusable by local sawmills.

Because he works alone, and his wife, who is a teacher, and daughter are away from home all day, Tom has to be extremely careful at every step in his work. His large studio is "essentially fireproof," he states; all flammable materials are stored outside in an old refrigerator that he keeps locked so that children or animals can't get into it. The studio has an air-circulating fan to disperse the fumes from his welding, and Tom often opens one or both of the floor-to-ceiling doors at each end of the studio for added ventilation. He has a phone extension in each studio, and the family has a code system for ringing the house in case of emergency.

The most dangerous part of his work on large sculptures involves climbing and intricate maneuvers in high places, yet Tom does not wear a harness or other special gear. "You do take risks," he acknowledges. "I'm very tuned in to it; when I'm doing something potentially dangerous, all my senses are geared to it. Since I'm alone here, the only person making noise is me, and any noise I'm not making means trouble. People can scare the hell out of me—if I'm in that situation and my wife walks in the shop, just the sound of her open-ing the door can catch me off guard. So I'm listening and looking at all those things at once. I've got to con-centrate here—if I have to worry about falling in addition to getting crushed, it's too much."

Tom shapes his sculptures with the same kind of attention and con-centration that has kept him safe throughout all the construction and risky work. He estimates that major pieces can take six to twelve months, and small sculptures absorb two to four months of his time. "I don't figure things in man-hours," the sculptor notes, "because I can't do anything in a man-hour. I count in weeks or months."

Gibbs spent nearly two years pre-paring for a one-man show of smaller works (each weighing about fifty pounds) in a Chicago gallery last year. He also had a large piece installed at the city's downtown Apparel Mart in conjunction with the show; this enabled potential buyers and groups that commission civic sculptures to see the full range of his work. Since 1970, Tom has been a finalist in a number of major civic-art competi-

tions, although he has not yet won such a contest. Several of his large works have been purchased by schools and municipal governments in Iowa, and the state arts council sponsored a touring show of his major pieces.

Although Tom finds that his profession is "a hair lonely" at times, he regards the proximity of his home and work as essential. "I think that's important to my work, because my life is a kind of totality here. I don't think in terms of working at odd times—it's just my life. There aren't any odd hours, because all hours are alike."

Gibbs is also an advocate of more worksteads and cottage industries, "because they are economically and socially practical," he asserts. "There's a tremendous potential here—how many people have woodworking shops in their basements that they never use? In most countries, that's a business—you know, you get a washer and dryer and you hang out a shingle; then you hire two employees. So there's tremendous potential for produc-

tivity, and it's all wrapped up in people's basements and garages."

The larger implications of this potential are likewise clear to Tom Gibbs: "Concentration of industry is a form of weakness—a couple of big boys make a couple of mistakes, and everybody's out of a job. You're not as vulnerable when you're wide-based, and you also have a population that is self-motivated, self-determined, and independent in attitude. You're going to have more people who are self-priming, and that's good for the society."

The Allens, The Brays & Victoria Stefani
NURSERY OPERATORS

"I really hate the waste. I think it's foolish if two people live next door to each other and you both have a canner that costs seventy dollars, and you've both got a lawn-mower. There's so much to do on this planet besides spend money and resources."

The viewpoint Barbara Bray expresses is central to the operating philosophy of the Creekside Nursery in Santa Rosa, California. Barbara, her husband, Jim, and their children, Michael and Molly, live and work there with nursery owner Jim Allen, his wife, Linda, and their son, Timothy, and Victoria Stefani. The nursery's eight acres contain some one hundred thousand trees and shrubs, a greenhouse for sprouting and care of delicate plants, a newly built barn that serves as the nursery office, and three small houses at the back of the property. The nursery is located on a long, narrow piece of land, so the Allens, Brays, and Stefani get from the office to home or to outlying growing areas on bicycles. Barbara and Linda, who only work part-time in the business, often carry their infants in backpacks while helping customers.

"We're practical ecologists, I guess you'd say," Jim Allen observes. "We're doing everything organically here now; we don't use weed killer, and we're getting away from a fertilizer that turned out to be partly organic and partly hype." One service for which Creekside is well known is providing live Christmas trees. "We don't like to cut trees—it's something that really pains us," Jim Bray notes. "It's more work for us to dig

them, but it's really satisfying to have people come in and say, 'Well, I used to have cut trees, but I don't want to do that anymore.' "

Jim Bray has worked at the nursery with the Allens and Victoria for several years, but he and Barbara moved there only recently. "Saving resources was definitely one of the things we considered when we thought about moving here," Barbara states. "We figured we'd be saving fifty to seventy dollars a month on gas alone." Jim adds that he also wanted to have a garden, which he has planted at the back of the nurs-

ery. "I don't want to be as dependent on my car," he says. "Even when you go to the market, you're using gas, and the food you buy has been shipped from someplace else, and that whole process consumes a lot of energy. So I feel that whatever I can produce within my own space is going to help all along the line."

The convenience of being close to work also attracts these five people, and they have established a healthy balance between joint activities and individual privacy. "I think we all respect each other's privacy as far as our living space is concerned," Jim

Jim and Barbara Bray and family; Timothy, Linda, and Jim Allen; Victoria Stefani; and part-time employee.

Bray notes, "but it's nice to have some people around you." Victoria observes that "We don't do a lot of social things together because we have enough time with each other at work." Linda and Barbara also state that they have been careful not to assume they "have an easy mark for a babysitter" in one of the other women living at the nursery.

All five worksteaders agree that they are not looking for a communal living situation. They maintain private residences and prepare meals separately, but some larger projects are shared, such as a garden and small orchard, a chicken coop, and such conveniences as a washing machine and a hot tub. The group has anticipated potential problems and discussed them, and they keep separate their relationships as employer–employees and landlord–tenants. As Linda Allen observes, "It's like a little community. We all have our work and our income and lots of our food right here."

The only rule that the group has had to make for the workstead is that the nursery's front gate must always be closed. This is done for the children's safety and because customers will ignore a "Closed" sign if there's even the tiniest crack in the gate. For these five people, the closed gate also creates a feeling of unity. "I love living here and having Jim working here," Linda notes. "It integrates work and home life—there isn't a separation." All four parents express their pleasure in being near their children, and all of them prefer to spend their free time on the nursery grounds. Victoria recalls: "I got feedback from people that they'd hate to live where they work when I told them I was going to do this. But it's been wonderful. At times I don't go out for a week."

Jim Allen's assessment aptly expresses the group's pleasure in their workstead: "The days when we're closed, it's like paradise here. I fantasize about living here for a year without going anywhere."

Bob Dunbar **&** Sally Dunbar
WRITER **&** PAINTER

"I feel the family roots very strongly. I like the idea of living where four generations of Dunbars have lived. Now it's six."

Ten years ago, Bob Dunbar went directly to his roots by purchasing the barn in Nobleboro, Maine, that his great-great-great grandfather built in 1796. He named the saltbox barn "Dunbairn," a play-on-words that incorporates the Scottish word *bairn,* which means "descendants."

As the name implies, the Dunbar sense of history and of family is thriving in this workstead.

The historical setting is ideal for both Bob, a freelance writer, and his wife, Sally, a painter, because they are surrounded by the subject matter they love. Dunbairn serves as a year-round gallery for Sally's work and as a warm-weather extension of the Dunbar family's living quarters beneath the barn.

Originally, the barn was intended as a vacation and retirement home, but after four summers of camping in the old structure, Bob and Sally and their two children decided to make it their permanent home and work place. Rather than changing any part of the barn's interior, in which the original hemlock beams and boards are still intact, the Dunbars had a six-room addition built into the sloping hill behind the barn,

and then had the barn moved back sixty feet onto the new foundation and living quarters.

"They actually moved the barn after all our furniture was in it," Sally recalls. "I couldn't believe we'd put everything we owned into a building that was about to be moved, but the van that brought our things from Illinois couldn't wait, and bad weather had kept the building mover from working on schedule. It took three days to move the barn, but nothing was damaged—one of our cats stayed in it the whole time."

The new living quarters and foundation for the barn were designed by Chicago architect Al Drogosz, who saw the building only once. He did not have detailed measurements of the structure but was able to calculate the barn's dimensions by studying photographs. His key to the scale of the building was a coffee pot in one photo—Drogosz used its exact measurements (which he knew because the Dunbars had it with them in Chicago) to determine the relative size of everything else.

The architect's fee for designing the addition was almost as unusual as his coffee-pot measurements: he did the job in exchange for two of Sally Dunbar's large paintings. Much of the Dunbars' commerce, in fact, has been in the form of barter. Sally has traded paintings for piano lessons and other services, and there's a sign in the Dunbairn gallery suggesting to customers that they barter for the art there.

The historic barn is an especially appropriate showplace for Sally's paintings, for many of them are connected with Dunbar family history, and others are inspired by the writing of Thoreau, who was a cousin of the family. One ancestor, Jesse Dunbar, Jr., was a surveyor in the area and kept a journal of the places he surveyed from 1811 to 1847. Bob and Sally have this journal, and Sally has found and painted many of the sites where Jesse worked. In addition to their Dunbairn home, Sally's paintings have been shown in the Chicago Art Institute and several galleries in

health column for the local newspaper, and one of his recent books is a young people's science-fiction novel, *Into Jupiter's World.*

Unlike Sally, who paints in the barn loft or outdoors, Bob spends his working hours in a cozy office that he planned as part of the family's living quarters. This room has its own electric heater, so if the rest of the family is out during the day, Bob shuts the door and heats only that room against the snowy and damp

Maine winter. "It's quite a different life, working at home," Bob says. "I never have to leave my office—or want to, because I enjoy it here."

Although the Dunbars are certain that their move from a Chicago suburb to a Maine coastal town was right for them, the transition has been difficult economically. "Our income dropped about two-thirds when we first moved here," Bob recalls. "There have been some hard moments, and we wondered a few times if we

Illinois and Maine, and one of her paintings was recently on the cover of *Maine Life* magazine.

Much of Bob Dunbar's research and writing also focuses on New England history, and he is active in Maine's Scottish society and president of the local historical society. His writing projects cover a wide spectrum, however; he writes a

should move back to the city." Bob left a well-paid job with a medical public-relations firm in Chicago, and it has taken several years for him to establish ample contacts as an independent writer. To help keep the family going, Sally taught school for the first three years the Dunbars lived in Nobleboro, devoting only her summers and weekends to painting. As the Dunbairn gallery has become better known, Sally's art sales (and trades) have augmented the family

budget significantly.

Dunbairn also has become known for community events. The local historical society meets there, and the gallery has been host to an archaeological exhibit, "Maine's First 10,000 Years," and student art shows. Two theatrical groups have performed in the barn, too—an old-time medicine show and a touring Shakespeare company, both of which drew more than a hundred people.

The Dunbars' choice of small-town life and their interest in living close to history have not made them feel at all isolated. As Bob puts it, "We didn't move here to escape from the world. We just like the idea of more peaceful surroundings in which to do our work and enjoy our family." Sally Dunbar agrees: "I feel very much at home in the barn, very secure. This place gives off good feelings from the past. It's almost as if the barn made us come to it."

Katharine Whiteside Taylor
JUNGIAN THERAPIST

"I think that it is a part of education to let people see how you live. Your way of life is very much an expression of your personality, and if you have a way of life that you feel people might like to emulate, then why not show it? Two people have said to me recently, 'You know, I feel as though I'm embraced as soon as I come into this house.'"

Katharine Whiteside Taylor has been a teacher and counselor for more than fifty years, and for the last fifteen she has used all of her home in that work. Since returning to San Francisco from the East Coast in 1966, Katharine has specialized in Jungian psychology, teaching "image seminars" and classes for as many as thirty people in her spacious upstairs studio and counseling individual clients in a small first-floor room. In addition, she occasionally hosts discussion groups, which meet in her living room.

Each of these areas has a special purpose, Taylor explains. "I find for work with patients you need a small, enclosed room. You're trying to become very intimate with one person, and that is enabled by a small space. I think of it as a kind of crucible where transformation takes place."

The counseling room is indeed small, with two facing chairs, a bed in one corner, a wall of bookshelves, and a small window; client and counselor sit almost knee-to-knee. The therapist points out that she takes notes only when she is beginning to work with a client, because she believes that note-taking interferes with the intimacy of the space and the process. "I don't want

anything between us—I want to be very close and get the real feeling of the person, which is just as important as what they say."

By contrast, the large upstairs room where Katharine teaches and writes is expansive and full of day-light. "When I planned the upstairs, which was built five years ago, I was thinking about what I like. I wanted a sense of space—a place to meditate and to relax—and this gives it to me." Taylor designed the exposed-beam, high-ceilinged room with the help of an experienced builder; they planned the space to utilize windows that were there from a previous owner's remodeling of the hillside cottage.

Just as the close quarters down-stairs are suited to therapy, Katharine

finds that the space and light of the studio are ideal for her teaching. "It has a very relaxing, soothing, and uplifting atmosphere," she explains. "I think that the students and people who come here feel at ease very quickly. That's what I want, because better thinking comes when you're relaxed; you're more receptive. I think that what comes out of the classes and image groups here is more effective because I'm doing them at home."

Because she teaches Jungian thought, many students have likened Katharine's studio to the tower that Carl Jung built as his work place. Taylor makes light of this analogy, although she graciously permits the comparison. "It's a loose inspiration, you could say. Jung did build a tower, but it was round and of old stones and entirely different. Mine is not like that—but, like Jung, I did want to go up."

The tower studio is also where Katharine writes each morning. During the last three years, she has reduced her counseling load from twenty-five to seven clients per week so that she can work on her autobiography and revise a book she wrote many years ago, *Parents and Children Learn Together*. Because writing is a consuming task for her, Taylor says, "I find it's hard to switch and focus on another person." She has solved the problem of divided

attention by establishing a daily routine that includes writing in the early morning and seeing patients in the afternoon: "In between, I always have a good lunch and a good nap; then I can go again—it clears my mind." Her daily routine also includes a morning warm-up of dancing, which she finds the most enjoyable way to exercise.

One drawback to her home-based career, Katharine relates, is that she doesn't have a secretary. Her modest living standards and her preference for solitude don't make that feasible, although she does have a typist come in for a few hours each week. Another concern in regularly using her home for classes is that she must keep it tidy. "I'm not such a neat person," Taylor confides, "and I'm likely to be leaving things here and there." She does have help with housework, though. A student attending college nearby lives in a small apartment in the basement of Katharine's home and helps clean the cottage in exchange for rent and a small stipend.

The integration of her home and her work far outweighs any inconveniences for Katharine Whiteside Taylor. "I feel that the atmosphere here is a very important one," the therapist says. "It begins with the sign over my front door, which one of my students made for me. It says *Témenos,* which is a Greek word that means 'a place reserved for transformation.' There's another sign I'm going to make to put over the door of the therapy room; it comes from the temple of Aesculapius, the mythical god of medicine, and it says, 'Come in well, go out better.' "

James Kobak
MAGAZINE DOCTOR

"Ten years ago, when I left a large accounting firm, I said, 'Gee, this has been a lot of fun. Now I'll go do something else.' Back then it was memos, meetings, and travel—my best trip was New Zealand and back in seventy-two hours. This is just the opposite."

James Kobak and his wife, Hope, have created a double workstead. Their Darien, Connecticut, home is the headquarters for James B. Kobak, Inc., and their Manhattan apartment is the city office where they see most clients. Jim Kobak specializes in magazine creation and reorganization, and he is well known for his work as a "magazine doctor" and for his trademark of casual dress, which usually features short pants. Kobak also works as a consultant for book publishers and is co-owner of a business that creates computerized publishing models.

When he began his independent business a decade ago, Kobak intended to work only in Darien. "I said, 'This is going to be great—I can have people come up to see me.' But I found that when people came to see me, they never went home. They would come up for a ten o'clock meeting and expect to stay for lunch. I decided that in my business I can't do that, so that's why we rented an apartment in New York. There I can move them in and out quickly."

For the past three years, Hope has worked as an editorial consultant for the firm, and Jim and Hope each spend several days in the city. "We go in like gypsies every time we go," Jim says. "I have five briefcases and Hope has two, and we often trundle food in with us, because it's so much higher in New York. The real problem we have is deciding what to take with us—it takes us an hour to figure that out."

The Manhattan office is an elegant apartment in the Olympic Tower, a midtown high-rise. Originally, Jim reports, "we got a space in the Algonquin, which we thought was camp for the publishing business. But then we found that the rooms were so small we went stir crazy, so we went to the Drake Hotel, and they rebuilt that, and we came here." One assistant works in this office—Mary Anne

Holley, a business and publishing specialist who plans to start a magazine with her husband when financing can be arranged for the venture.

The Kobaks find that their fortieth-floor city workstead has an unanticipated advantage for keeping clients to their scheduled appointment times. "Right outside our window is the Newsweek Building, which has a huge clock on it," Jim notes. "When people come here, they sit on the couch and look directly at that clock. So they can't miss the time—and that helps keep them moving along."

Because clients rarely visit the Darien workstead, Hope and Jim do most of their quiet work and phoning from there. The office's location at one end of the house affords privacy for the living quarters, and the space is large enough to provide offices for Jim and Hope and their two secretaries. A copier, computer terminal, and several filing cabinets are neatly fitted into the office space, but much of the voluminous paper generated by this word-oriented business has been relegated to "dead" files in the garage and a former chicken coop.

Although Jim still travels often—about half as much as in his former job—he does much of his consulting work by telephone. "I must make two hundred fifty calls a week," he states. "And it's the funniest thing—you treasure the telephone hours. I don't go out to lunch anymore; those hours are too valuable. When I go to the West Coast I have the time of my life, because I can get a day's work done between

six and nine in the morning." Hope would like to lessen the number of business calls, however: "I've clamped down on that—clients can call all day, but they don't have to call at night."

Most support services are available in the Darien area, so a suburban workstead hasn't caused any hardship in that sense. Kobak notes that the local printing and duplicating shops don't have quite the range of services of city suppliers, but messengers are available for rush jobs that require a New York firm. The Darien–Manhattan messengers also are kept busy ferrying in files or books to Jim or Hope at one office or the other. All the firm's mail comes to the Darien home, and the volume is such that the mailman routinely brings the load to the house instead of putting it in the couple's mailbox. "My friends leave things in the mailbox for me," Hope says, "and I never find them because we never open it. The mailman drives right up to the door."

Despite the travel and the packing of papers to take from one office to the other, Jim and Hope Kobak find their two-part workstead ideal. "I get to see Jim," Hope says, "and I can keep up with my friends and my bridge group." Jim agrees: "We do the really important things—she has to play bridge every Monday, and I try to be in Darien every Thursday night in winter, because that's when my bowling team meets. And in summer, it's golf—I have a regular game here called the Kobak Open."

His contemplation of the "really important things" leads Jim Kobak

to one other observation: "This is the way people are going to work in the future—with all the communications devices, there's no question about it. We all don't have to be next to each other all day every day. I often give speeches on the subject of 'What is the future going to be?' which is easy, because nobody knows anything about it, including me. The only thing I know is that the changes will happen more slowly than we expect."

Sheldon Baumrind ORTHODONTIST & Joan Finton ARTIST

"I don't think in terms of the other way—a regular office—because this seems so reasonable and appropriate. I think if I were put in a situation where I couldn't easily drift downstairs and do my work, I'd be very restless."

Dr. Sheldon Baumrind is able to "drift downstairs" from his home to the spacious office that houses his orthodontics practice and research lab. The setting for his workstead is an 1875 Victorian house that is among the oldest of its genre in the San Francisco area. The building recently won an award for the contemporary interior remodeling that complements its historic exterior.

In addition to its design merits, this century-old Victorian is a fully utilized building. Dr. Baumrind purchased it in 1967 with the intention of setting up his office on the first floor and renting the two second-floor units. The downstairs required extensive renovation to fit the needs of the orthodontist's work; this involved installing many new cabinets, adding wiring and plumbing, and even removing a bearing wall to open up the waiting room. "The nature of orthodontics is that you have lots of kids flooding in late in the afternoon," Dr. Baumrind notes, "so you've got to have a reasonably large reception area. When I laid out the office, I planned a big one for here."

The waiting area is indeed large and prominent, encompassing the bay-windowed front room of the downstairs. The grasscloth walls are decorated with children's art, and a built-in aquarium and shelves of reading materials offer diversion for the patients. A feeling of openness and light is maintained in the office and treatment areas, as well; low standing walls and cabinets permit natural light to flow through the space, and the placement of the dental couches allows patients to see what's going on around them. This

spacious treatment room is distinctly more pleasant than the cubicles found in most commercial medical buildings. Two other small rooms are located on the first floor—the back office, where bookkeeping and billing functions take place, and the research area, which Dr. Baumrind and one or two assistants use most

evenings and weekends.

The coming and going of employees and patients could have caused a traffic or parking problem in the Baumrinds' neighborhood, which is a mixture of residences and small commercial buildings. For this reason, local officials required that Dr. Baumrind add a paved driveway and parking area on his property, which obliterated the building's yard. He was given a use permit for his office after complying with this requirement and adding commercial-type insulation to one outside wall of his house. "I think the city planning people here were far-sighted, and they were glad to have someone buy the building who wasn't going to tear it down," Dr. Baumrind recalls. Because of the area's zoning, a buyer could have razed the house and replaced it with a multiple-unit apartment building. This was never the orthodontist's intention, however: "We tried to save the character of the building and the area as much as

we could. I think we've been pretty successful."

Renovation of the upper floor of the Victorian came several years after the orthodontist began working in the building. Dr. Baumrind and his wife, the artist Joan Finton, and Joan's daughter Nora have lived in the house for the last three years. "The thought of combining my office and home only came to me after I remarried about five years ago," the orthodontist says. "It stemmed from the fact that from 1968 to 1973 I lived downstairs in the back. I didn't really think of it as living with my work; it was just sort of an economy. But when I moved away, I realized that I really did miss the proximity to my working position, because I do work a lot in the evenings."

The move from a rented home in a more isolated hill section of the area has suited all three members of the family. Nora, a high school student, "likes to be in the midst of the bustle of the city," Dr. Baumrind notes.

Joan Finton also likes "being down near people," and her studio space was a primary element in the remodeling plan: "As soon as we started thinking about what shape this would take," she says, "the studio was part of it."

Adding an artist's studio without changing the exterior of the house was a major design challenge for architect Peter Behn. The ceiling was lowered in the back part of the upstairs living room, creating a loft studio under the roof while keeping the spacious bay window and high ceiling at the front of the room. Skylights were added to provide the essential natural light for the studio, and some additional lighting was installed for evening work. One result of opening up the attic area and adding the studio loft was that the customary hiding place for heating ducts wasn't available. This problem was solved by hanging the heating unit from the low ceiling in the central hallway, so that it became a design feature of the space and opened into the large living room, which requires the most heat.

Both the studio and living areas are decorated with Joan's work. Two large paintings of the magnolia tree in front of the house are central features of the living room, and Joan's paintings and graphic works brighten most walls upstairs. The art's varied colors, set against the clean lines of white walls, and several seating clusters give this home an inviting sense of comfort.

Although the living areas are not large, Joan says, "It's a beautiful, spacious-looking apartment. We're certainly satisfied—it's a very happy trade-off." The studio space also adds to the living area when necessary; it has been used as sleeping quarters for Nora's friends, and a jazz combo has played there during parties.

Of course, the studio is principally the center of Joan's work. She teaches painting there three nights a week, and one morning each week a group of women artists meets in the studio to sketch a model whose fee they share. "There's a lot of conversation and kidding around," Joan notes, "and we all get the advantage of sharing the cost of a model." The group numbers eight, which is the limit of the studio's space. At Christmas, the artists sell the work from these morning sessions—the paintings and drawings are displayed in Dr. Baumrind's office and reception area.

Before moving to California, Joan had taught in conventional school settings in New York, and she finds teaching in her home preferable: "The advantages are that you have everything at hand that you need— I've got supplies, I can keep files of visual aids, and I have my books, so I can always run downstairs and get a book to illustrate something in my teaching." Her pupils also enjoy the comforts of a residential setting. "Everybody loves the house," Joan points out. "The students who come here generally are not prepared to see such a beautiful modern house when they come in."

Like her husband, Joan Finton enjoys being close to her work. "I don't

have to go traveling—I'm here. And I can run upstairs right after dinner for a class." In fact, students often arrive as the family is finishing supper, "and that may be a little less formal than some people expect, but they get used to it quickly."

The informality and comfort characteristic of this family are evident in their easy flow from a leisurely meal to an evening's work. On the nights when Joan teaches, Dr. Baumrind works in his lab, where he makes computer-assisted three-dimensional measurements of minute changes in such processes as facial growth. The techniques Dr. Baumrind has developed and refined are being used in his work with the University of California in radiology, orthodontics, and orthopedics. The variety of his projects is a principal reason why Sheldon Baumrind likes to work where he lives: "There are a lot of fragments here, and if you're going to work evenings or at odd times, it's very nice to have all your things about you, and not to realize suddenly that you left something somewhere else."

Joan Finton has a similar perspective on her workstead: "Teaching at night is very nice, because it's ordinarily a time when I'd just collapse, maybe in front of a television, and probably not do my own work. So teaching is both a way to work in art and it's also a social occasion. And it happens that Shelly likes to work at night doing his research. It's a way we can each just go in different directions in the same house and even have time sometimes to go out for a drink afterward."

Ralph Day
RESEARCH PHYSICIST

"It becomes a tremendous convenience to live with your experimentation. Many critical testing programs are relatively ineffective because the quitting whistle blows, terminating an incompleted test prematurely. There is an important cost and effectiveness advantage of living in the laboratory."

Ralph Day has conducted twenty years of experiments in his workstead laboratory, a 150-year-old home in Maumee, Ohio, where he lives with his wife, Harriet. A research physicist whose specialties are glass and heat conductivity, he recently has developed a new type of thermally insulated window, which has a replaceable balloon filled with a desiccant to absorb the moisture and vapors that cause fogging between the panes of these double windows. In the course of developing and testing his new designs, Day rebuilt the frames for his own windows in his basement workshop, tested various desiccant substances in his small laboratory, and installed the new windows in their original places in his house. "My windows are noticed all the time," Ralph says, "to the same extent that we watch birds flitting around our windows 'all the time.' "

This new project follows from Ralph's study of heat transfer and because, he says, "I got tired of hauling the storm windows around each winter." Day's penchant for handling all phases of a project is characteristic of his research career. When he began his workstead consulting in 1961, a nearby manufacturing firm hired him to make 100 measure-

ments of heat conductivity in small samples of a ceramic product. The physicist readily agreed, only to discover that the testing methods and the equipment available for this work were not accurate enough for the job. So he spent nine months refining the test process and then built all of the testing equipment. "I made all the high-vacuum equipment myself because I had no idea if I was ever going to get any of my money out of this," Day reports. "And I had time, rather than money, to invest."

Ultimately this series of tests was successful for both the client and the scientist. The manufacturer had Day test samples of competitors' products in his neutral laboratory, and his results showed that his client's product was consistently good. Ralph patented the special measurement device he developed for the project and has received royalties and made consulting trips to several nations as a result of the invention.

The scientist's inventiveness is reflected in his home laboratory, too. Many of the measurements in his heat-conductivity testing require the use of very sensitive instruments that can be affected by any nearby motion. To alleviate this problem in the lab, Day mounted his vacuum pump on a concrete slab that rests on a bed of old tennis balls—"one of the best vibration dampers there is," he notes. Today such accommodations would not be necessary, Day adds, because newer electronic measuring devices are not so sensitive to exterior forces.

The tendency of his older equipment to vibrate from outside movements, such as traffic going by the house, was one problem that Ralph's workstead location helped him solve. "Even with the tennis balls under the pump, I had a problem with vibration—I could actually see the effect of trucks going by a block away. So I made some of my measurements in the middle of the night to avoid this movement."

A home location also permits Day to work on other projects or to simply relax while he is waiting for an experiment to reach completion. "Most of the measurements take a long time to stabilize," the physicist says, "so when somebody comes to the house, if I am all set up, I can leave the laboratory and wait for stability while I'm having a drink. Then I'll go back and take a few readings."

One drawback of an independent laboratory is that the scientist must find sources of materials. "Big companies have specialists who locate these things," he says. "I have to do

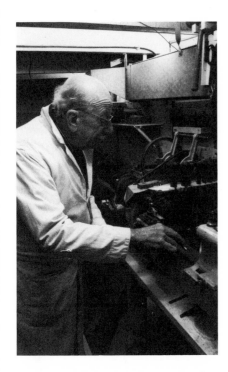

it all myself." But Day has found this task pleasurable as he canvasses junkyards and rummage sales for much of the material from which he fashions instruments. "A research lab has to have the most miscellaneous things in it, and accumulating things to make this lab was a lot of fun," he observes. "I have the ability now to do just about anything. I can build anything in the way of instrumentation, and I have an oxygen tank, silver-soldering tools, and a brazing torch. Tremendous capability exists in this house. There are a lot of people who boast that 'Now I'm going to stop working.' I'd be very distressed if I had to stop working. I like to be at home, and I like not being bored at home, so I continue to look for projects, and I'll be this way as long as I live."

Charles Fox
GUITAR MAKER & Joanne Fox
PUPPET MAKER

"We've followed each other's progress, and I think we all have been an active part of each other's careers. We're each pursuing our own professional goals, but we have the kind of relationship that allows, encourages, and demands that we serve as trusted counsel for each other."

Guitar maker Charles Fox shares in a cross-country relationship that involves four separate workstead businesses. The professional and personal friendship he describes is among Charles and his wife, Joanne,

Dewitt Jones & Babs Kavanaugh
FILMMAKER & TOY MAKER

whose home is South Strafford, Vermont, and Dewitt Jones and his wife, Babs Kavanaugh, who live in Bolinas, California. Joanne and Dewitt were high school classmates, and they re-established contact about ten years ago; since then, their two families have grown close and their individual careers have prospered with mutual advice and support.

In addition to making guitars, Charles Fox supervises workshops in that craft, and Joanne Fox makes puppets that are sold in galleries and museum gift shops. They have lived on their sixteen-acre Vermont property for a decade and tried farming for two years before establishing their workstead businesses. Their enterprises have grown so steadily that both of the Foxes now have employees, which they believe is particularly valuable in their rural location. "There are a lot of people in this area and not much work available," Charles states, "so it's good that Joanne and I can supply some jobs."

Charles's guitar-making business includes two full-time employees; one makes the custom-designed instruments that Charles markets to individual customers and a few shops, and the other teaches the year-round workshops in the craft. The guitar workshop and teaching facilities are housed in a small building behind the Foxes' home, and students are provided dorm-style housing in the "cupcake" yurts the couple built before constructing their spacious log home, which Joanne designed. In addition, Charles conducts research in both acoustic

and electric guitar design and performance, experimenting with new body styles and sound effects for the instruments. He also has an apprentice, who works without pay in exchange for Charles's instruction; the man hopes to be hired by Charles when his year's apprenticeship is completed.

Joanne's puppet-making business recently has grown too large for her home studio. Although she still does much of the design and handwork on the puppets at home, Joanne has rented a small studio in the town

where her daughters go to school, so that she now combines the twenty-minute drive to school with several hours of work with her full-time employee in the new annex. She also has hired two local independent contractors to make clothing for her puppets from designs and fabrics she supplies.

The workstead businesses of Dewitt Jones and Babs Kavanaugh are located across the country from the Foxes' Vermont workstead—in Bolinas, California. Because Dewitt's film and photographic career re-

quires that he travel at times, he is especially glad to have his studio across a field from his home. Originally Dewitt and Babs had both of their work places in one small studio that Dewitt built behind their house, but both his career and Babs's toy business have grown too large for them to share a space. So the couple bought a second home, and Dewitt now uses their first house as his editing lab, office, and photo center, while Babs has taken over the original studio.

While Dewitt's work is largely a one-person discipline, Babs has nine part-time employees in her workstead toy studio. All are from the Bolinas community, and most are mothers who also are practicing artists and craftspeople in their own right. Babs feels that supplying jobs to other women is one positive offshoot of her business. "I think it does

a lot for the women's self-esteem to be earning and contributing," she observes. "There also is a social situation that's been created in my studio. The thing about the nuclear family is that it's very isolated, and this gives women an opportunity to come together not just to watch the children. The element of conversation, of sharing problems and ideas, is something I see as very important."

The two couples' workstead connection has been especially significant for Babs and Joanne. Seven years ago, when Dewitt spent a year in Vermont making the film *Robert Frost's New England,* Babs traveled with him and took a hiatus from her toy-design business. "My mind was on my business back home when I got to Vermont," Babs recalls. "It was really exciting to meet Joanne, because I hadn't been expecting that anything creative could happen dur-

ing that time. Joanne had a lot of tools as well as lots of ideas. I got to tap into her creative energy, which was a saving grace for me."

At that time, Joanne did not have a business going; she was working with fibers and soft sculpture in her home studio and, with Charles, was rearing their two daughters, Danielle and Meera. "I've been doing artwork since I was twelve," Joanne says. "When the children were small, I was just drawing, but at the time Babs came out, I was doing soft sculpture. While she was in Vermont, Babs and I would get together and make things—just experiment and try out new techniques that we'd seen or heard about. And that was our way of entertaining ourselves."

During the time that the two women worked together, Joanne developed a technique for making

the puppets that have become her business. "Once we discovered this technique for the soft sculpture, I began making them with the excuse that I was making them for the children. But after a while it became clear that I was much more interested in it than the children were, and Babs and Charles both inspired me and gave me the support that I needed to market my work. I didn't know the first thing about how to do it, but Babs is a walking encyclopedia, so I'd just pump her when we were together, or I'd call her up with questions."

Babs points out that the benefits of working together were mutual, for she was introduced to Joanne's knowledge of design, which added a new dimension to her own work. "And," Babs adds, "we worked with the children, which was a valuable experience for me and is something I've done with Brian since he was very young. For a small child, it's really important for him to feel competent, to feel like he has control of what he does, and that happens by having him participate in what I'm doing."

The Foxes' daughters, who are now teenagers, have helped fashion Joanne's puppets from the beginning, and Joanne agrees with Babs about the value of this work to children: "It's problem solving, really, that we're involved with all day long. And the children see it being done successfully—and they do it successfully, on their level. Instead of always being told that you have to wait until you're an adult before you'll be able to handle those things, they're feeling like you can move

each step when you're ready."

Charles Fox notes that the family's working together also strengthens their relationship. "It's not just that the kids get to understand what their parents do, but the fact that they do it with us means that we have a common vocabulary. Usually parents establish a vocabulary between themselves and kids by somehow talking down to them. Instead, through the vocabulary of handwork and manipulating materials, which our kids really do, we all understand that that's who the Foxes are. We have a sense of who we are and a shared experience."

Dewitt Jones adds to this workstead perspective from his experience with his three-year-old son: "I know it's sometimes irritating for me to try to find the line where my business stops and my life starts," he states. "But the good side is that the studio and the house have a much easier flow than I ever did with my father. He left at eight in the morning and I didn't see him until seven o'clock at night, and I had no idea what he did. So there's a much closer tie here in that regard. Brian said to me the other day, 'You're a photographer, Daddy.' I didn't teach him that; he came up with it himself—and that's neat."

The two families visit one another at least once a year and keep in touch with each other's work more often. The similarity in the development and success of Joanne's and Babs's worksteads has meant that the women continue to share business information and advice. Charles points out, too, that the four friends "save the big questions for each other—like the direction we should take in our careers. We sort of feel the others looking over our shoulder." Dewitt echoes this thought: "I don't respond to Charles's guitars in terms of whether the bridge is in the right place, but in terms of the grand philosophy of the vision he has, and whether he's on track with that. He bounces that off me as somebody who also treats his life as his art."

Tom Scheibal
ANTIQUES DEALER & INNKEEPER

"If I'd kept a 9-to-5 job, I never would've seen my son, Quinn. I've been able to spend a lot more time with him because I'm right here—my house is my office. It's kind of a juggling act, but it sure is fun."

Tom Scheibal's "juggling act" includes a country inn and antiques business, a city warehouse that was his original workstead home, and his single-parent role for a seven-year-old son. Tom may be one of the few people who have two worksteads, one in the city and one in the country. Both involve his antiques collecting, refinishing, and selling activities, and for six years the city warehouse was his and Quinn's residence.

That original home and business was the U.S. Antique Depot, located in a hundred-year-old sack warehouse next to a malt factory in San Francisco. Tom stored and worked on antiques in the huge downstairs

space, kept files and a phone in an old boxcar he brought into the warehouse, and lived in the elegantly furnished upstairs loft with Quinn and their dog, Conrad.

Scheibal found this arrangement ideal for taking care of his young son while keeping the business going. "When we first moved into the warehouse, Quinn was quite young; my business hours were his nap hours," Tom recalls. As Quinn has grown older, Scheibal notes, he has been able to participate in the business in small ways. "He's seeing me in action, and he wants to refinish furniture or hammer a nail or sweep up the shop," Tom says.

The home-based business also exposes his son to a variety of people, Scheibal points out: "I think he's going to have an incredibly rich childhood, because he comes in contact with lots of people—he's very sociable. And his playmates love to

come over here and climb around this old furniture."

The warehouse workstead also provided a great business advantage for an antiques store—its location. "Unlike a grocery store or a regular business, where people go and expect to buy a pound of coffee or a shirt," Tom observes, "an antiques store is a discovery place. At first I was working in a basement, and then in this funky old warehouse. When people walked in here, they'd think, 'I can't believe this, it's like a dream—I've got to buy something. I've discovered something.' And that's basically what my business in that location was based on—the discovery theory."

Unfortunately, the city warehouse was sold recently, and it will be torn down to make way for new housing. Scheibal had anticipated this, however, and about three years ago he put his profits toward the purchase of an old tavern building in St. Helena, in California's Napa Valley wine country. Tom maintained the city business during the week, then had a friend tend the shop on weekends while he and Quinn traveled to the country site, where he worked on remodeling the building into a downstairs antiques shop and an upstairs inn. About a year ago, Tom and Quinn moved to St. Helena permanently, at first staying in the future guest rooms and subsequently in a precut log cabin Tom had built behind the inn.

The collection of antiques that had been stored in the warehouse and used in the loft now provides most of the furnishings for the five

guest rooms in the Bale Mill Inn and Antiques. Each room is named for a literary figure and decorated appropriately for that person: there's the Jack London room, with Indian relics, rustic furniture, and a wolf under the bed; Emily Dickenson, all wicker and New England crisp lace; Captain Quinn, a mythical sea captain, with nautical gear all around; Teddy Roosevelt, a formal, stuffy presidential suite; and Ernest Hemingway, with its Key West furnishings and a stuffed parrot—by far the most popular room.

Scheibal operates the inn and the retail antiques business himself, with the help of one employee who cleans the inn's rooms. He serves a light breakfast to the guests each morning, then opens the shop during the day. His new location is a different kind of discovery for many customers—they often come to look at antiques and end up making reservations to stay after Tom shows them the upstairs rooms.

The inn's location on an acre of land at the edge of miles of open country also suits Tom as a place for his son to grow up. "I prefer this small town to the city for Quinn," he states. "He has a wonderful mixture of friends here, and behind our house is about twenty-five miles of wilderness area where he and his friends can play. I have a big old bell from a firehouse that I installed in a tower on top of the cabin—it's the only way I can call Quinn in from outdoors. You can hear it all across the valley—that's quite a difference from the city."

Tom Scheibal has turned the necessity of caring for his son and earning a living into a workstead vocation that has prospered in two places. "It's everybody's dream," he says, "to really enjoy what you do and where you live. For me, they're the same thing. It's sort of like living on a boat; you're a fisherman, and right below you is your catch."

Othmar & Mildred Klem
FURNITURE MANUFACTURERS

"My wife and I, we own the plant; the boys work for us. It's strictly a family organization."

Othmar Klem's family organization has grown from a 720-square-foot cabinet shop, where he and his wife made cedar chests in the evenings, to a 43,000-square-foot furniture plant that makes four hundred items per week. The family home is next to the original shop and is now somewhat overwhelmed by the acre of factory and warehouse fanning out behind it. Othmar and Mildred Klem have five sons and three daughters, all of whom have worked in the plant at some time; at present, sons Jim and Bill work full-time for the family business, and Dan and John work part-time. One of their daughters, Susie, spent several months working in the plant last winter, which is her landscaping profession's off season.

The Klems' furniture business is located in the small town of St. Anthony, Indiana, and is one of the area's larger employers. Some seventy men and women staff the company, which specializes in solid-wood furnishings for hotels and motels. Since 1973, the firm has produced furniture for twenty thousand guest rooms.

Othmar Klem was born in the house across the street from his present home, and he and Mildred first lived in rented rooms in a house next to the cabinet shop, which they subsequently bought. Klem has made woodworking his career since the 1930s, when he found a job as a

carpenter's assistant. "In the early thirties, we had a tomato cannery up here—that's where I really got the liking of wood, when I built the cannery with a carpenter," he recalls. Klem began to do fine woodworking as a hobby and gained general experience by working in a nearby furniture plant for twelve years. He and Mildred paid for their house by selling cedar chests, and in 1948 Othmar went into the cabinet-making business on his own when a three-phase power line needed to run the necessary machinery was extended to St. Anthony.

Having the shop—and then the plant—in the yard has been a real convenience, Othmar notes. "Sometimes we'd take the children to the shop in the evenings," he says, "and in bad weather we never have to miss any work time." Mildred always has worked in the business and kept the household going, and she recalls working around infants' schedules and the plant's hours. "When we were making cedar chests," she states, "we'd cut one out in the evening, and the next day I'd glue it. I had my children trained—they slept until eleven thirty or twelve o'clock, so in the morning I could work in the shop and go back to the house in time to fix lunch and take care of them." Mildred still has family to

feed each day, for Othmar, Bill, and Jim usually have a snack during their nine o'clock break and lunch at noontime in the family home. As Bill points out, "It's the best restaurant in town."

On several occasions the proximity of living and work areas has saved the business from disaster. One evening the family was preparing to go to church and Othmar was ready before the others. He went out to check the shop, Mildred reports, and found it on fire: "He came flying out of that shop and over to the well and threw a five-gallon bucket of water right on the fire. Then he yelled for me to call help." St. Anthony had no fire department at that time, but all the neighbors came and helped douse the small blaze. "Half an hour later," Othmar says, "and it would've been gone." Besides this fire threat, the Klems' sons have thwarted at least two robbery attempts.

Most of the benefits of a family workstead are more routine, however. Bill observes that he learned respect for quality from his parents' work: "One thing is that working here and living here got us kids involved in understanding wood and manufacturing. We knew what quality was, because we always had it shown to us." He adds that the extra money he and his brothers and sisters earned in the plant, usually during high school summers, helped to pay for their college educations. Dan Klem recently took up one of his parents' original crafts—he made cedar chests for Christmas gifts last year with help from his father and his brother John.

Othmar Klem is pleased with his children's interest in woodworking and is especially proud of how far the family business has come: "What's brought us to where we are today is that we always try to make a good piece of furniture, to put good workmanship and finish on it, and stand behind our product." Mildred adds, "My whole aim was that my next piece of furniture would be better than the one I just finished. I still try to do that."

Patty Gleeson
RECORDING STUDIO MANAGER

"I think working and living in the same place creates a feeling that's very beneficial to your business. I know for me, if I hadn't been living in this building, I wouldn't have put the amount of energy and the hours into the studio. I've worked incredible hours, but I think you have to do that to make a business successful."

Patty Gleeson's business is Different Fur Recording, a state-of-the-art audio studio that is housed in a turn-of-the-century warehouse in San Francisco. Patty and her husband, Patrick, share both the living quarters and their careers with Different Fur; she manages the studio and business, and he composes and records music on the elaborate synthesizer that occupies part of the second floor of the building, where he is currently completing the movie soundtrack for *The Plague Dogs*. In a recent remodeling of their studio, the Gleesons installed forty-eight-track recording, along with thirty-two-track digital capability; they've also added a live echo chamber for their clients' use, as well as a SYMPTE system that synchronizes all the recording devices and is specially geared for film and video soundtracks.

Patty Gleeson has lived in the three-story studio building since 1970, originally as part of a group of fourteen people. After the group dispersed, Patty and Pat Gleeson and one other couple decided to improve the building's living spaces and begin a recording business on a small scale. The two families handled the construction, sound engineering, and business operations themselves; then the Gleesons' partners decided to move on, and Patty took the reins of the business when Pat went on tour as a synthesizer player.

Although the studio originally was tailored to Pat's musical needs, Patty has built the business and the facilities into a successful commercial venture. "We've always been ahead of the market," Patty says, "even though we may be smaller. We're making the studio profitable, number one, because we live here, and because clients have always gotten more for their money at Different Fur. We've always put our money back into the business; we're constantly improving. You can't stay

in one place in the audio business—you have to keep making it better, or you're losing ground."

One of the initial problems in establishing a professional recording studio in the wood-frame building was sound leakage—both inside and out. "This is a frame building where you can't use cement as a damper, so isolation for sound is a problem," Patty states. "We've solved it to the point that no sound from outside gets in, and very little sound from inside gets out. Bass vibrations from the studio do carry within the building, though."

Living with the sound vibrations and the comings and goings of clients can be draining, Patty acknowledges. "There's never a weekend when the building isn't full of people recording, and some groups work all night or very late. Their friends are here, and often they're doing demo tapes, so they're playing as loud as they can. For me, because I'm not a musician, it's sometimes like living on top of a store."

The Gleesons do live at the top of the building, in a handsomely appointed space they designed and built themselves. A large back room, once slated to be Patty's painting studio (she studied art before taking over the recording business), is devoted to equipment for the new echo chamber and a small sleeping loft for guests. The open front half of the floor combines the kitchen and living room, above which there is a sleeping platform. Because storage space is at a premium in this open arrangement, Patty's clothing is stored in a closet just under the roof, which is

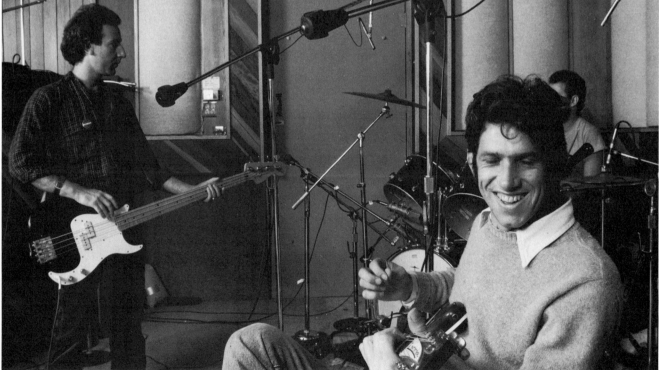

reached by an extension of the winding staircase leading from the second to the third floor. The location of the bathroom at the head of the stairs occasionally causes an embarrassing moment, Patty notes; a few times a house guest or client has come to the living quarters after an all-night recording session, only to bump into a sleepy Pat or Patty in the process of getting dressed.

Such inadvertent invasions of the couple's living quarters are likely to be diminished by the newest addition to the studio's facilities, however. A lounge and kitchen have been added on the second floor of the building, so that clients can relax between sessions or even stay overnight there. A sauna and shower also were added and are separated from the lounge by a lovely curving wooden wall. This spacious new area for clients is one of Patty's innovations in the firm. "Once I took over the business," she explains, "to make it creative for me, I wanted to expand the horizons. I'm very concerned about the comfort of our clients, and I'm involved in making the best use of this building for them and for us."

When the lounge area was built, the first-floor entryway also was remodeled. Patty's office was moved to the second floor, and a reception area and food counter were installed downstairs. "For our professional image, I want this place to look good," Patty says, "and I realize that each group in the studio wants a certain kind of food, and that it provides a certain professional touch if I go and get that food for them. And it keeps me in touch with the clients—

it's a nurturing situation."

Patty Gleeson believes that this combination of personal touches and professional service is the core of Different Fur's appeal: "I think it's true that because we live here, we really keep it as nice as we would our home. Lots of musicians come here from Europe or the East Coast because they find it so private—they can stay here. For some people, it's like their own New York loft."

Lloyd Prasuhn
VETERINARIAN

"I do it all because I think the important thing is dedication—to the community and your clients. You have to provide a community service; that was the whole idea in building this hospital."

Lloyd Prasuhn, D.V.M., is the director and chief veterinarian of Lake Shore Animal Hospital in Chicago. Since 1967 he has lived and worked in the 15,000-square-foot facility, maintaining an animal-care practice that began in this part of the city in the 1890s. "This is the second oldest practice in Chicago," Dr. Prasuhn points out. "It was originally established for the carriage trade, to treat the horses that used to pull delivery wagons."

Dr. Prasuhn doesn't see any horses in his urban practice today, but his hospital is equipped and staffed to give the most sophisticated care possible to small animals. "We have about fifty thousand clients," Lloyd says, "and on a busy day I'll see as many as a hundred animals." The veterinarian is assisted by a staff of twenty-six technicians, handlers, and other personnel, including one other full-time vet who specializes in birds. In addition, Dr. Prasuhn works regularly with a group of consultants who make periodic visits to the hospital (and stay in the second-floor guest quarters) or are called in for special cases.

The veterinarian previously had his office in a storefront and his home was close by. But when he had the opportunity to build the new hospital, he felt strongly that he should live in the building. "My wife, Mary, and I designed the hospital and

our apartment upstairs," Lloyd reports. "When we lived two blocks from the old office, I'd have to go there two or three times a night. And you lose a lot of time just coming and going. You have to get the car out, or sometimes a car would be parked in front of the driveway, or it was bad weather—you just can't cope with coming and going and running this kind of hospital." Now

Dr. Prasuhn simply gets in the elevator to have quick access to home, office, and all the hospital facilities.

Another way that the workstead vet keeps in touch with his patients is by television monitors, which link the animal cages, intensive-care room, surgical areas, and the Prasuhns' apartment. Another monitor permits clients to see their pets in a special "visiting room," and the

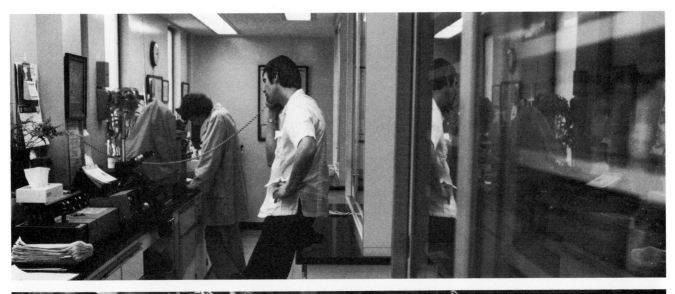

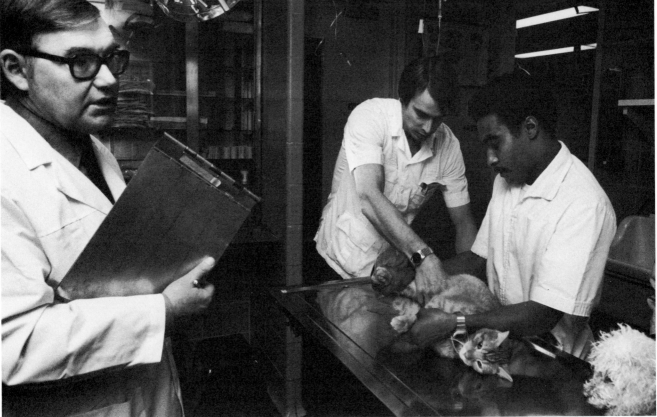

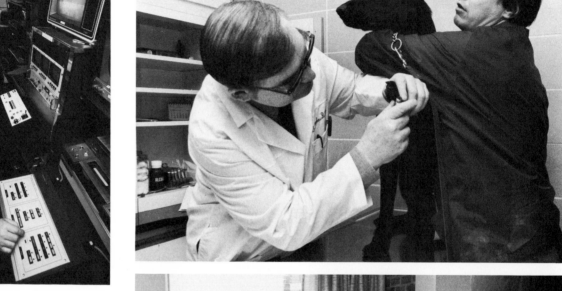

front and side entrances of the hospital have built-in cameras that are monitored at the front desk so that the person on night duty can detect any potential intruders.

Recently, Lloyd set up a video studio in one room of the guest apartment that shares the second floor with his own living quarters. "I want to give something back to my profession," he says, "so I'm going to start making educational films for clients and other veterinarians to use." He also has videotaped much of the surgery done in the hospital, and his collection of slides of interesting cases numbers at least forty thousand, the vet estimates.

As in many workstead environments, Dr. Prasuhn's professional materials have appropriated some of his living space. The slide collection and many articles and books occupy a large portion of one of the two bedrooms in the couple's apartment, and Mary Prasuhn's office, where she handles most administrative duties for the hospital, consumes the remainder of that room. Lloyd and Mary don't feel cramped in their spacious quarters, "but we'd like some more storage space for our own things," the veterinarian admits.

Some of the storage problem also is caused by materials for a closely related interest of the Prasuhns—the nonprofit Lake Shore Animal Foundation, which is an animal adoption center. Lloyd estimates that the service places three hundred to five hundred homeless dogs and cats each year, and he helps to fund the operation by offering donation premiums, such as T-shirts and record albums, that are stored in the family apartment. Usually the homeless animals are brought in by neighbors or clients; they are treated if necessary and kept in the boarding area of the hospital until a home can be found for them. Sometimes, too, the boarders become residents of the vet's apartment—currently, four cats live there.

Because of his dedication to both of these animal-care centers, Dr. Prasuhn spends most of his time in this one building. "The only real problem that I have," he says, "is that it's just plain too confining. You get to feeling like you're really cooped up. I've been in the hospital maybe as long as a week at a time without going out."

When he and Mary do go out, Lloyd carries a radio-paging device so that his staff or answering service can notify him in case of an emergency. Eating in a variety of local restaurants is a regular source of relaxation for the couple. "That's what we do for recreation a lot of nights," Lloyd notes. "We like to find authentic places, and Chicago has all kinds. I probably have seventy-five restaurant owners as clients, which is one advantage of my business."

But the major advantage of a workstead to Dr. Prasuhn is a practical one: "The main thing is that someone must be present all the time to cope with developing mechanical problems or emergency problems. Like one Saturday night, I heard this squeak as I went out to the garage, and I knew it was a mechanical problem. It turned out that a fan belt was slipping on a motor, so we had to take care of that, because if something happens to our air flow, we're in serious trouble."

The more common situation is the late night or early morning phone call or knock at the door. "I came downstairs about six o'clock one morning," Lloyd recalls, "and I heard a car honking outside. I opened the garage door and this guy drove right in. The man's sheepdog had been hit by a car, and he had brought it right here. If I hadn't been here, the dog would've died."

Goodman Building
ARTISTS COMMUNITY

"The vitality here is because each of us is a politician and an artist and a member of a community. It's a large creative organism; the building and the life that goes on inside it are functionally and integrally related. There's a synergy here."

Painter Martha Senger is a resident of the Goodman Building in San Francisco, a community of artists, writers, musicians, and actors who are committed to sharing resources, ideas, and rents. Their group workstead takes an uncommon form, because each member of the Goodman Building community lives and works in what is essentially a hotel room, rather than a self-contained apartment or loft. In this historic thirty-

seven-room building, the twenty-five residents share kitchens and bathrooms and work cooperatively at such chores as building maintenance, elementary repairs, and cleaning of the common areas.

The wood-frame structure was built in 1869 as a two-story residence and was sold in 1900 to a family named Goodman, who rebuilt the dwelling after the 1906 earthquake to include four stories, with five storefronts at ground level and a fourth-floor studio that has a magnificent skylight. In 1907, photographer H. Pierre Smith rented this studio, becoming the first of many artists to live and work in the Goodman Building. As the building itself

and the concept of shared housing for artists have enjoyed a recent renaissance, Goodman Building residents and supporters have done extensive research to document a seven-decade tradition of artist-residents here; the list includes dozens of poets, painters, dancers, and in the 1960s, singer Janis Joplin.

Painter Byron Hunt has lived in the same room in the hotel for sixteen years. The walls of his room are covered with his art, and he enjoys working in the high-ceilinged, well-lighted space. "There's a real community here," Hunt observes; "and many talented people. I have what I need to work." Like many other supporters of this arts community, Hunt remembers similar residential buildings in San Francisco that were torn down for commercial purposes; in fact, he once lived in the last of them, Montgomery Block, which was razed to accommodate the pyramid-shaped high-rise that houses Transamerica Corporation.

Byron Hunt, Martha Senger, and their fellow residents have been working for the last eight years to save the Goodman Building from a similar fate. The struggle over the building began in 1973, when the city's redevelopment agency appropriated the property, which is located at the far fringes of downtown San Francisco. The agency planned to tear down the somewhat dilapidated structure and permit development of housing in some other form. With help from a coalition of arts supporters and sympathetic citizens, however, the Goodman residents made a successful case

for the building's designation as a city landmark and then for its listing in the National Register of Historic Places. This effectively halted the building's demolition, but the official historic protection was awarded only to the façade of the structure. The group's petition for recognition of the cultural value inherent in its continuous use by artists did not meet historic-landmark criteria, so

the interior of the building can be altered so long as the façade is not disturbed.

The plot thickens at this point, because Goodman Building tenants began to withhold their rents—ordinarily paid to the redevelopment agency—when the agency declined to make any repairs in the hotel. The diverted rents were used to fix a leaky roof, install some better wiring

and heating, and make other temporary improvements, with much of the labor supplied by the worksteaders. The artists also secured a $15,000 grant from a state historic preservation group to repair the leaking skylight, but the redevelopment agency, as the building's legal owner, rejected that money and a subsequent $20,000 grant that residents had arranged. Ultimately the

Goodman Building residents hoped to buy their collective home from the city for its declared value of $196,000—a sum they could have raised by securing a number of small grants.

The redevelopment agency, however, chose to support a commercial developer's plan for renovation of the Goodman Building. The plan calls for conversion of the structure into one- and two-room apartments, each with a kitchen and bath. Funding for this million-dollar project was to be awarded under the federal Housing and Urban Development department's "Section 8" program, which provides rehabilitation funds and rent subsidies for low-cost housing. The main stipulation for such funding is that each unit in the building contain a kitchen and bath, so the present hotel arrangement doesn't qualify. Critics of the commercial redevelopment plan point out that rent subsidies for the apartments to be created would total $3.2 million over twenty years, because low-income residents who qualify for such housing are required to pay only one-fourth of their income in rent, with subsidies making up the difference. Thus, some $4 million of public money would be spent to house the same number of people living there now without any subsidy.

The Goodman Building's residents have countered the commercial de-

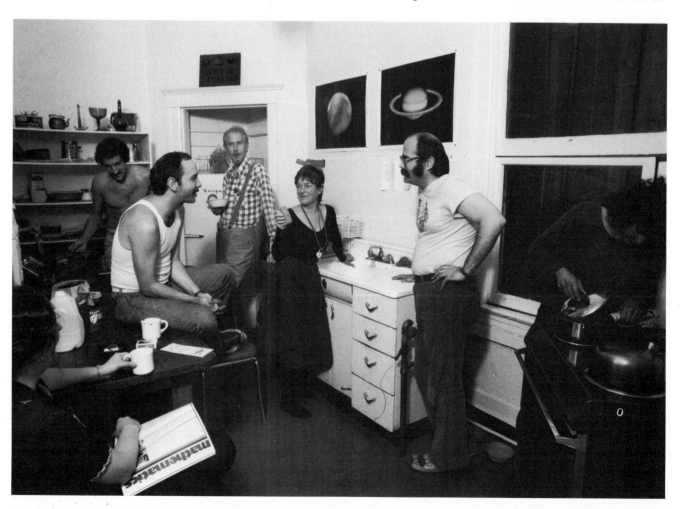

and theatre in the storefronts of the building. In these public spaces, classes are offered to neighborhood residents, fine printing equipment is made available to people who don't live in the building, and dramas are performed by Goodman residents and other artists.

Martha Senger, who has been particularly active in the campaign to save this building for its cultural value to both the community and its low-income residents, believes that any city would be poorer without its artists. "I came to San Francisco because I'm an addict for intense experience," she notes. "That's why artists come to cities—for those complex forces, the excitement of activity. It's going to affect art if artists can't afford to live in cities anymore; I'd starve aesthetically and intellectually if I weren't in the city."

Senger does not need an apartment with kitchen and bath to work as an artist; rather, she and her colleagues thrive on the balance of privacy and community that they have maintained in the Goodman Building: "It's pragmatic, economical, and efficient. To me, what's important as an ecological issue—not just physical ecology, but the human creative ecology—is that you tie efficiency and housing by using the simplest and best materials and spaces. Why put anything in here that's going to decrease the efficiency of the art function of the community? I know that by going to a community kitchen to fix my lunch, I'm going to run into somebody and we'll share ideas—something bigger will happen."

veloper's proposal with a detailed restoration plan of their own, which would cost about half a million dollars to bring the building up to the city's legal hotel code and would include purchase of the structure by the tenants. This plan, which involves the "sweat equity" labor of the residents, is supported by a number of urban housing experts and city planners, among them state housing director I. Donald Terner. "The Goodman tenants are inviting government to save its very scarce subsidy dollars for projects where there are absolutely no alternatives," Terner states. "The tenants are telling us that they'll roll up their sleeves and do the job themselves—without public funds. To prevent this effort not only wastes taxpayers' money, but it stops other projects for truly needy families which can't go ahead without these funds."

After eight years of hearings, petitions, protests, and appeals, the Goodman Building political football is still in the air. The artists have found wide community support and have continued to improve their residence in the modest ways possible with limited funds. They have made contributions to the larger community, too, by painting the handsome façade of the hotel and by establishing a gallery, printing shop,

HOW

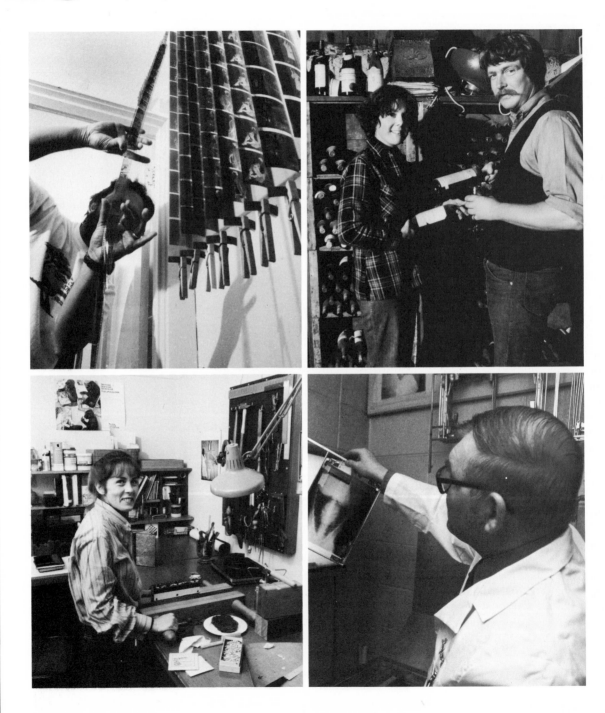

W e've all had the impulses that inspire worksteaders—wanting to stay home, to quit the commute, to work for ourselves. Yet even with these desires and the example of established worksteads, it may seem a quantum leap from a conventional office or factory job to a home-based career. In fact, the change is a big one, and it touches every aspect of a person's life. That's why the major portion of this book is devoted to the essential considerations for establishing a workstead livelihood.

Being in business may be familiar territory to many people who become worksteaders, but operating from home makes a difference. A residential location often affects the availability of supplies, services, and even customers, and can influence the "business image" as well. Employees are working in an office that is combined with a home, which creates special responsibilities for both workers and employers.

The spaces occupied by workstead businesses are tremendously varied and highly individual. Some of the many innovative uses of space in worksteads are presented here, and architect John Edwards has contributed an article that explains and illustrates three ways to create a home office.

Money, of course, is a primary consideration in starting a workstead business. Many worksteaders are seeking and managing funds for the first time and must take extra care not to mingle household and business finances. One definite advantage of a workstead is that it provides home-office tax deductions, which accountant Nancy Feiner details in her article here.

Worksteaders also must contend with certain legalities, such as the zoning and building-code regulations of their communities. Some cities have responded to the revival of structures for both working and living by amending their laws to accommodate home businesses. In most places, however, the legal ground still must be plowed by individual worksteaders.

And, perhaps most important, worksteaders don't have a built-in separation between their work and their domestic lives. They always know that the desk or work table is just a few yards away, so they must establish priorities—finding ways to close a mental door that will allow them to enjoy leisure time and to get back to work when necessary.

These are the principal concerns in beginning a workstead career. Though some of these considerations are not exclusive to home-based occupations, the information and ideas presented in the following pages focus specifically on working and living in the same place.

Doing Business

All worksteads have one characteristic in common: they must survive as businesses, or the work isn't going to support the living. Although their occupations and environments vary widely, worksteaders generally share another characteristic—they are responsible for all aspects of being in business, from inspiration, production, and selling to licking stamps. And few of us are natural whizzes at all these tasks. Fortunately, all manner of experienced people have shared their wisdom—in books, magazines, newsletters, newspaper columns, and seminars—so there are ample tools available for learning what's essential in the operation of any workstead business.

The topics discussed in this section are especially relevant to work-live businesses. Being at home—in either a residential area or an industrial zone—usually separates the worksteader from many of the support systems and daily contacts available to those who work in office buildings or other commercial locations. Yet there are some advantages to being set apart from other businesses. This distance, for example, forces the worksteader to plan his activities and develop an organized operation—something many businesspeople never master.

Getting the hang of doing business is particularly important for the large percentage of worksteaders who are in the precarious position of starting a new business. The lawyer who quits a firm for a solo practice, for instance, suddenly finds herself making appointments and sending out bills in addition to consulting with clients and arguing cases. Thus, she must budget her time to balance the earning activities—hours spent on clients' actual business—against the nonpaying but necessary support functions. Or she has to find assistance with the support work so that she can devote her time to income-producing tasks and to finding new clients. Perhaps a more common example is a craftsperson who spends all his time and money making handsome furniture without arranging for sales outlets to recover his expenses, earn income, and thus stay in business. In situations like these, when worksteaders are novices at running a business, it is essential that they identify their special work needs and choose from among the alternatives that will keep them from joining that unenvied majority of new ventures that fail.

Advice from Experts

Particularly for worksteaders who are starting an independent business or tackling a new career, some friendly advice probably will be useful. In fact, there's so much advice for sale that determining what's valuable may add to the confusion of embarking on a new enterprise. There are some reliable and relatively inexpensive resources, however.

Publications

In addition to the many books on all aspects of doing business (some of which are cited in the bibliography), several other types of publications offer specific help for worksteaders. The most useful and directly applicable of these is Bank of America's *Small Business Reporter,* a series of large-format pamphlets that cover individual occupations and a variety of business functions. One issue of this series—"Steps to Starting a Business"—should be basic to any workstead library. This twenty-page booklet contains an excellent summary of the obligatory considerations in establishing a business, and its contents should serve as a quick review for even the most seasoned businessperson. (A list of other pamphlets in this series and ordering information are included in "Resources.")

The federal Small Business Administration (SBA) is well known for its loan programs, but it also publishes a number of pamphlets and bibliographies pertaining to most areas of business. These publications are generally less comprehensive than Bank of America's series, but they contain useful reading lists and often cover more narrowly defined topics. Other sources of low-cost and generally reliable information are the federal bookstores in major cities and the clearinghouse for federal publications in Pueblo, Colorado (see "Resources" for address).

Two new magazines also provide much useful information for worksteaders, and some of their articles are specific to home-based businesses. These publications are *In Business,* which is geared to small businesses and contains a regular column by Bernard Kamoroff, the author of *Small-Time Operator* (an excellent book about business in general); and *Venture,* a journal devoted to entrepreneurs that contains pertinent data on sources of capital, taxes, and legislation affecting new ventures.

Personal Contact

A surprising number of people in established businesses and careers are quite willing to help newcomers. These experts often enjoy giving assistance or advice, and their valuable experience may be tapped most effectively if the worksteader is organized in asking for information and considerate of the businessperson's time. Unless such a person is in a highly secretive or competitive occupation, he often will encourage newcomers in his field, greeting them as signs of a healthy business climate rather than fierce competitors. Independent publisher Malcolm Margolin, for example, readily shares his experience with aspiring book producers, because he believes that a growing number of flourishing independents will serve to increase the attention paid to these firms by the publishing industry at large. He also finds that the casual atmosphere of a home office makes him accessible to would-be authors and publishers: "One of my major businesses," Margolin notes, "is free consulting."

One immediate source of advice may be friends or acquaintances in a particular field. Other sources of advisors are trade associations or professional societies in the worksteader's field, as well as librarians and local college professors. In-person assistance also is offered by the Service Corps of Retired Executives (SCORE), which can be contacted through the SBA. These consultants have broad business experience and the objectivity that may have escaped a struggling worksteader, and they frequently work for expenses only.

There also is a group of small businesses—many of them worksteads—that provides its own business consultant to members. The Briarpatch, a loose association of enterprises that de-emphasize traditional profit and competition motives and promote a concept of "simple living and right livelihood," is concentrated in the San Francisco area but has chapters in many other cities. In addition to sharing certain business practices, such as open books to anyone interested, Briarpatch members contribute a small sum each month toward the salary of an experienced business advisor. This person consults with members, visits their work places, and will even participate in the running of a firm for a brief period if the owner is ill or needs temporary assistance. The Briarpatch chapters meet regularly—both to share ideas and to socialize—and a volunteer staff in San Francisco publishes a quarterly newsletter called the *Briarpatch Review* (see "Resources").

Classes

Community colleges and university extensions offer a variety of practical classes in business operations and specific vocations. These courses are frequently free or low-cost and usually are taught by professionals in a particular field. Many commercial consultants also offer seminars and workshops, usually on weekends; these classes tend to be more high-powered and high-priced, and their value often varies according to the composition of the group attending the event. In any class, whether a one-day seminar or a semester course, it's most useful to ask specific questions, insist on detailed answers, and help to shape the content to the students' needs.

Contacts and Customers

Being located away from established commercial areas can make finding work and finding customers difficult. This isolation is one of the trade-offs for a workstead's comfort and convenience. Nevertheless, with some effort and ingenuity, worksteaders can locate customers who prefer a neighborhood business to an enterprise that operates in a shopping center or skyscraper.

Basically, home-based businesses fall into three categories: those that produce items for sale, those that sell items produced elsewhere, and those that offer personal services. Worksteaders who make or sell products must find customers to buy these products; those who

Deborah McDermott:

Going to the Source

Negative cutting is an uncommon skill and a highly specialized part of filmmaking. It is also the profession that Deborah McDermott learned by asking an expert if she could watch him work. Deborah was in college studying film arts at the time, and other filmmakers told her to look up Kenji Yamamoto if she wanted to learn about negative cutting. "He didn't know me," Deborah recalls. "I just called him up. He's real accessible—because of his location— so there are a lot of people who go in there looking for jobs. When I called him, he said, 'I'm working on a film right now. If you'd like to come over and watch, you're welcome to.' So I did, and I sat there trying not to move—not to disturb the dust— and watched him work."

Deborah was sincerely interested in learning this exacting craft of cutting the film negative to the editor's specifications —the last and most critical step in preparing a film for reproduction. Because the "neg cutter" handles the original film stock, one mistake can ruin an expensive, and perhaps irreplaceable, scene. McDermott

understood this, and her questions and persistent interest led Yamamoto to share his knowledge with her. "He started by giving me one job—four years ago—and then referring more and more clients to me," Deborah states. The timing of her informal apprenticeship was fortunate for both Kenji and Deborah, because Yamamoto had just begun making a transition from neg cutter to film editor.

While she was still in school, McDermott used part of her student loan to set up her negative-cutting room at home. This area must be free of dust and air movement, so

she sealed the windows and even added a piece of rubber to cover the crack at the bottom of the door. Her first jobs there were other students' films: "I would beg people who were doing film projects to let me cut their negatives as practice."

McDermott's skill and care were evident in this early work, and now she enjoys a solid professional reputation and has more requests for work than she can handle. In fact, presently she is teaching her craft to a student, so that he can share the negative cutting as she adds film-editing projects to her work.

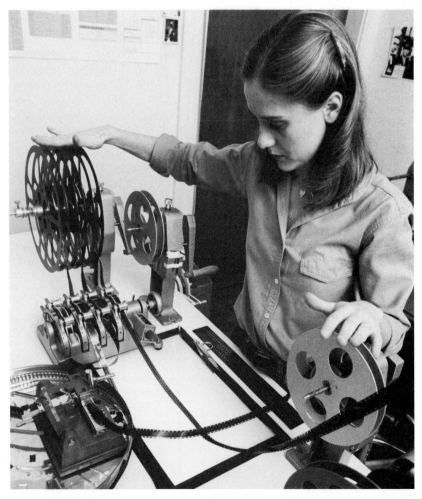

offer personal services must develop contacts that will lead to clients. Regardless of their product or service, however, all worksteaders must make certain efforts to establish their businesses and build a clientele in order to survive successfully.

Business Image

As with all enterprises, a business image must be established for public credibility and internal morale. Particularly for worksteaders who are newly independent, there may be a temptation to acquire the most elaborate trappings of "being in business"—expensive machinery, fancy furniture, gold-embossed stationery. These accoutrements are nice, of course, but most worksteads are shoestring operations that would be quickly bankrupted by such a load of nonessential expenses. Rather, workstead businesspeople must make a realistic assessment of their immediate needs, which could be defined as the ability to deliver a competent product or service in a professional manner. This might translate, for example, into inexpensive business cards and stationery, a rubber stamp for invoices, and regular trips to the local copy center instead of a home copier.

Whatever the particulars, however, the worksteader is in control of her business image and can make it as sophisticated, flashy, or homespun as she wishes. It is important that she choose a way of defining herself as a member of the business world, and she should prepare for this choice by evaluating her goals, the market for her product or service, and the general climate of her competition. (If that sounds like a business textbook, it should—there's similar advice in every one.)

In creating an image, worksteaders have the advantage of using their unusual office settings to attract customers. A graphic designer whose workstead is a Victorian house, for instance, uses a drawing of the gingerbread-decorated structure on her stationery and business cards. And a rural publisher whose home and studio are located in a lovely agricultural valley puts out a newsletter called *Book Farm* and likes to describe his business as "the publishers in a cow pasture."

Advertising

If a business image is the professional "aura" for a workstead, advertising is its trumpet. And most worksteaders need to sound off for customers or clients to know about them, because they seldom are located in well-traveled commercial areas or launched behind the name of a celebrity. Although new businesses and small ones—which covers a high percentage of worksteads—can't afford high-priced exposure such as television or national magazines, the principles of mass advertising still can be applied to promoting them. An excellent digest of types of advertising and ad strategies is presented in "Advertising Small Business," a pamphlet in Bank of America's

Small Business Reporter series. The most basic point made in this report—and all other ad primers—is that every business enterprise should plan and budget for some advertising to acquaint potential customers with its services or products. Similarly, regardless of the type of product or service, the most successful advertising and marketing will be done by businesspeople who define precisely the markets for their products or services and who gear promotional efforts to those markets.

In addition to the traditional messages communicated in print or over the airwaves, advertising can be accomplished in other ways. For example, if a product or service has news value—a truly better mousetrap or a car that runs on water—the free publicity that results may establish it in the public's mind. A recent workstead example is the West Coast distributor of wind generators who constructed one of the huge towers and blade assemblies in his yard. He got plenty of media attention and some complaints from neighbors for this effort, but city officials couldn't find a regulation against this energy-saving device, so the generator is still in place. And the man got another round of publicity when he demonstrated the generator, showing how it provided electricity for the lights in his home office.

Advertising also can be employed on a very small scale. Notices on bulletin boards, signs placed in local stores, and similar inexpensive efforts can be effective for neighborhood businesses or services. One group of craftspeople who work and sell from home loan their work to local shops for display, in return for a sign that tells who made the items and how to purchase them. Another group of craftspersons and artists rents space in an antique shop at Christmas time and leaves information about their work in that shop during the rest of the year.

Networks

Word-of-mouth is the strongest source of work for many worksteaders, especially people in such professions as teaching, counseling, and catering. Architect Bob Hatfield notes that most of his business comes from people who've been referred by clients; Peggy Swan and Howard Levine have depended entirely on word-of-mouth to attract customers to their art workshops and guest house.

Groups that share a common interest likewise can provide useful contacts. Many occupations have informal "grapevines" that a worksteader might tap for sources of work or sales. The artisans guilds that thrived centuries ago have been revived in many places, and their members share customer lists, among other assets. Leisure activity groups also can offer valuable contacts; an artist who specializes in drawing or painting birds, for example, might have his work reproduced in the local or state newsletter of the Audubon Society or similar association, which could lead to important commissions and sales.

Sometimes, too, participation in a group effort can lay the groundwork for a whole career. This was accountant Nancy Feiner's experience when she decided to support the fledgling *Women's Yellow Pages* by taking a paid listing in the book. At that time she was establishing an independent accounting service, and as a single parent and working woman, Nancy felt that her contribution to the publication would help two worthy causes—theirs and hers. Her investment has been returned manyfold; in the four years since the book was published, Feiner has built up a workstead accounting business with five hundred clients, most of whom found her through the source book.

Selling from Home

In many places, zoning regulations do not allow residences to be used as retail sales outlets (see "Legalities"). Some worksteaders discreetly ignore such regulations; others sell their products through shops or sales representatives elsewhere. Worksteaders who sell wholesale products are not affected by rules prohibiting retailing from home, but often they require large spaces, such as industrial buildings, to store and ship their merchandise.

Some worksteaders are able to use their homes for the preliminary steps to retail selling, however. Painter and ceramic sculptor Donna Billick displays much of her work in the living room of the large old farmhouse she shares with three other occupants. She does not invite customers to her home, but often has gallery owners visit her studio in one of the farm's outbuildings and then view her finished work in the more formal setting of the house. This process has worked successfully for Billick, who keeps these necessary interruptions to a minimum by inviting visitors only for specific, limited times.

If a workstead is zoned for commercial use, it can accommodate both living quarters and a retail business. Antiques dealer Tom Scheibal found such a building—a warehouse with a spacious loft, located in an industrial and commercial area. He used the large downstairs for storage and refinishing and lived in the three-room loft with his seven-year-old son. The loft's furnishings came from his stock of antiques, which led to Tom's occasionally selling the contents of his living room. "Most people didn't know I lived there," says Scheibal, "but sometimes I'd invite people upstairs. And if I ever wanted to sell anything in particular, I put it in the loft and sold it from there. They had to have it when they saw it upstairs—'because it's out of his own home.'"

Perhaps the most common method of workstead selling is by mail. Because mail service is so widely available, this type of business can be located almost anywhere. If the volume of mail is too great for regular delivery, the worksteader may have to pick up mail at a local post office. All mail-order selling is regulated by several agencies. The Federal Trade Commission prohibits unfair advertising and sets time limits on the filling of mail orders by businesses. The Food and

Deanna Jones:

Doorstep Sales

down the street comes to tell me about the birds she's seen on her walks." One satisfaction Jones finds in having a workstead occupation is this chance to meet neighbors. "The pottery is a good reason for starting a conversation—it's very tangible," she says. "And once you've talked about something, it's easy to be friends."

The entryway shelves of finished work also attract interest and sales: "Neighbors who need a gift will knock on the door. Lots of times it's people who've seen me working, and sometimes it's a customer who bought my work somewhere else and wants another piece."

The potting studio has become a family gathering place as well. Deanna's daughter, Hilary, often works with her mother, and at other times she plays on the trapeze that hangs from the garage ceiling. Deanna's husband, Ron, a teacher, spends much of his free time on writing projects, working

in a room at the back of the garage. "It's really nice to be downstairs and have Ron bring me a cup of coffee or Hilary come down and tell me she's going to do something," Deanna says. "Those are really important connections."

The sales from her home display and studio are not Deanna Jones's only outlet, though. She belongs to a seven-member potters guild and also participates in holiday sales with a group of diverse craftspeople. During one recent Christmas season this group of about forty artisans took over one member's home and yard for a sale, using every inch of the home's downstairs floor space and furnishings for displays in the "department-store-for-a-day." With publicity sent to participants' mailing lists and local businesses, this weekend event drew three thousand people and made more than $10,000 for its members.

Potter Deanna Jones uses her home as both work place and sales depot. Just inside the front door of her Mediterranean-style home (entry and garage on ground level, living quarters upstairs) is a row of shelves used to display her work. Narrow windows on either side of the door light this area and enable would-be customers to see the assortment of teapots, mugs, planters, vases, and other items, even if no one is home.

Most days Deanna is at home, however, and working in her ground-level studio, which is a converted corner of the garage. She opens the garage door to add light and air to the rather dark space, and this has brought both kibitzers and buyers. "It opens up the house," Deanna notes, "and people will just wander in and ask me what I'm doing. The little-old-lady bird watcher

Drug Administration regulates all sales of food, medicine, and cosmetics. The Postal Service establishes intricate regulations governing the types of items that can be mailed, package sizes, and much more. To ensure a good working relationship with postal officials, a worksteader should seek answers to specific questions before preparing a mailing. It also helps to deal with the same person on each visit to the post office, because many of the regulations are subject to personal interpretation by postal employees.

Employees and Other Forms of Help

Many worksteads are established by people who want to escape an organization and work independently. But as their workstead businesses prosper, these folks often end up creating and managing new organizations. Other worksteaders choose to remain one- or two-person operations, but frequently they need part-time or temporary help to keep up with the work load. Whatever the circumstances, however, hiring people to help with a business is a commitment that requires careful consideration.

The Option of Growth

Before hiring employees or signing agreements with independent contractors, a worksteader must determine that his business can sustain the growth and meet the costs of additional help. For example, an employee who comes to the workstead will need work space and furnishings; some equipment may have to be duplicated for the added person; and payroll costs can include Social Security taxes, unemployment insurance, and somebody's time for handling the increased paperwork that an employee entails. Hiring an independent contractor for specific tasks—performed in that person's home or work place—does not involve payroll or space, but it does commit a set portion of the business's resources to this work. Thus, the worksteader should have an accurate timetable for integrating any assistant's work and payment into his operation.

Even before determining what type of help is appropriate, it is important to evaluate the desirability of expanding the business. In many instances, growth is the only option for long-term survival. But just as often, the "quality of life" considerations that originally motivated a worksteader also may argue against growth. These equally valid needs probably will have to be reconciled more than once in a workstead's history.

Choosing Employees

Once a decision has been made to hire employees, another process of consideration begins. The new people (or person) should have skills—and preferably experience—that suit the tasks required; therefore, the workstead employer must define the precise duties

John Harris:
Mail-Order Mania

John Harris has created a workstead business that has its own mail-order persona. As Lloyd J. Harris, he is the author of *The Book of Garlic* and founder of a nationwide club for garlic fans called Lovers of the Stinking Rose. Harris keeps LSR members informed of the latest developments in garlic theory and practice through the mail with his newsletter *Garlic Times,* and much of the garlic news comes to him by the post as well.

"I'm a mail addict," Harris admits. "Every day, opening the mailbox—there could be money, there could be offers, there could be letters of praise or entertaining letters. Articles on garlic appear at a rate now that I can't keep up with, and one day I'll get twenty-five letters from Podunk with an article in them—and to me it's just a thrill."

Harris's garlic correspondence club has about two thousand members, with one or two new applications in each day's mail. *The Book of Garlic,* now in its third edition, has sold nearly twenty thousand copies, largely through mail orders. John notes that he began his garlic research as "serious fun" and that now "it verges on a cult, because garlic creates such fervor among people—it's not like an ordinary food. I get letters from all over the country, and lots of them are from people who have a reputation in their family or community for being garlic fanatics, and they're so glad to hear about this group, because now they can have an identity, an affiliation."

John Harris has provided these people with an affiliation, and the "serious fun" with this mail-order business also has been successful. So Harris has undertaken a new project—to compile and publish a guide to organizations that operate through the mail, ranging from outlandish correspondence clubs to societies with a poignant purpose. Harris has attached another variation of his name, L. John Harris, to this new book, which will be published by the firm in which he is a partner, Panjandrum/Aris Books of Los Angeles and Berkeley.

of potential employees and hire qualified people. Moreover, the employer should monitor her staff's efforts regularly, for the quality of their work can affect her professional reputation. Robin Davey of Glad Hand Designs, a kitchen accessories manufacturer, notes: "We've built our business on quality, and the only control we have is the paycheck. We could farm out work to sewing contractors, but the work doesn't come back the way we want it. So we have to screen and hire our own people."

After employees are hired, they can be most valuable to a workstead business if they are familiar with all aspects of the operation. Their motivation is likely to be high if they understand how their work fits into the larger business effort, and they can best represent the business if they know it well. This knowledge also can be particularly helpful in freeing a worksteader's time, because the assistants can answer inquiries and handle phone calls.

In some worksteads, too, the employees become temporary members of the household during business hours. If the workstead office is located in a home or loft, for instance, the business is likely to spill over into living areas, so the people working there should be sensitive to the privacy of the worksteader's family and—ideally—compatible with family members. The workstead employer can influence both employees' and nonworking family members' attitudes by clearly defining the work space and establishing protocol for the business.

One of the easiest solutions to finding employees for a workstead business—hiring a friend or relative—also can be the most difficult to manage. The employer–employee relationship is fuzzy in such instances, and the fact that the workstead is the center for both work and social activities can make personal relationships even more awkward if there are problems on the job. If friends or family members are employees, the worksteader can help to avoid misunderstandings by establishing a clearly professional relationship for all business activities.

Whoever the employees are, their presence adds a new layer of bureaucracy to the workstead. There are employee taxes (see "Finances"), insurance, possibly a retirement plan, and forms to file for each of these. Depending on the type of business and its size, the Occupational Safety and Health Administration (OSHA) may have applicable rules as well. The federal Department of Labor dictates the minimum wage for most businesses, and the IRS defines who is considered an employee (generally someone whose work is directed by the employer and for whom the employer provides tools and space).

Independent Contractors

It is often easier for a worksteader to use independent contractors than to hire employees. This outside assistance really must function as a separate business, however, or the independent contractor could

Christopher Weills:

Employees in a College Town

One unusual workstead business helps a lot of other worksteaders sell their work. *The Goodfellow Catalog of Wonderful Things,* the creation of editor Christopher Weills, presents a huge collection of items by individual craftspeople, most of whom work at home. Weills is also publisher of the *Goodfellow Review of Crafts,* a bimonthly newsletter that lists shows and events, profiles artisans, and reports news of interest to practicing craftspeople.

The center of this crafts-publishing activity is a Berkeley, California, garage that houses a computer typesetting machine, a few hundred boxes of files and photos, an assortment of contributors' work, and Christopher's living quarters. Weills's operation is definitely low-budget; for example, he rents time on the typesetting machine, often losing sleep while a client works there through the night. He uses the garage rent-free, compliments of his brother-in-law, and several of the staff members are volunteers. The newsletter editor is paid a small salary, and the rest of the staff—as many as a dozen workers—are students from the University of California.

Because the Goodfellow operation was set up as a nonprofit foundation (with help donated by a lawyer), Weills can hire students in the university's work-study program. "The work-study students have a fairly good wage—about five dollars an hour, I think. They're paid through the university and the state," Chris notes, "and the allocation we pay is 20 to 40 percent of their wages."

Although the students' schedules change each quarter, and there's a fair turnover from year to year, Weills is well satisfied with his comparatively small investment in wages. "The students have been extraordinary. They're disciplined, they rewrite my letters and correct the grammar, and through twelve years of school they've been taught to turn things in on time. So when they come to us, they understand about deadlines."

be defined by the IRS as an employee. Two workstead examples illustrate this distinction. Robin Davey of Glad Hand Designs has four employees who sew aprons, tea cozies, and other items under her supervision; one of these women works in Davey's workstead and the others work in their own homes on projects Robin gives them. Lisa Goldschmid, who designs and makes clothing, does much of her own sewing but takes all of her designs that require quilting to a family workstead that specializes in this intricate work. The family has many other clients besides Lisa and is clearly an independent contractor performing one specific task for Goldschmid. Both women are happy with their assistance: Robin prefers to train and supervise her employees (and is willing to do the paperwork for them), and Lisa gets expert specialty work whose cost she can incorporate into the prices of her high-fashion clothing.

Independent contractors can be found for most routine business tasks, such as clerical work, bookkeeping, and public relations. A number of home-based businesspeople do typing from telephone dictation; their services are especially popular with doctors, who often dictate case records after normal business hours. The advantage of phone dictation equipment, of course, is that input can be recorded at any time, and the person who transcribes the dictation also can work at his or her convenience.

Workstead businesses often work together in this way. Attorney George Hellyer, for instance, has a varying amount of legal paperwork, which occasionally includes a document or report that must be prepared very quickly. This secretarial work is done for him by a neighbor who formerly worked as a legal secretary and now has a home office. Because these two worksteaders live close to each other and have flexible schedules, Hellyer's routine and rush jobs have been completed to the satisfaction of both working professionals.

Another workstead mixes employees and independent contractors who all have home-based operations. Barbara Dean's Island Press has two part-time employees in addition to Dean and an independent publicist who works on contract for the firm. One employee's residence also serves as the branch office for this small publisher.

Finding Help

A worksteader's main considerations in seeking help are competence, cost, and convenience. We're all probably a little tight and a little lazy, but most of us would pay a bit more or look a bit harder for expert assistance. And, in many instances, that help is already in business, so that screening and hiring employees—and the extra time and expense this requires—may be unnecessary.

This is where professional and community networks are invaluable. For example, Robert Ueltzen, a retired telephone-company employee, serves as coordinator for a network of independent servicepeople whom he met through his phone company work.

David Burton:

Hiring His Successor

For the past seven years, David Burton has done a thriving business as a children's photographer. He converted the garage of his home to a darkroom and office, and added first one, and then a second, employee to help him print and mount the ten thousand pictures that parents order each year. His photographs also illustrate a number of publications, including a children's art book, *Don't Move the Muffin Tins.*

Like many workstead employers, David found that he had to take on more business to keep his employees busy. Ultimately, this increased work load led to an opportunity for one of his assistants, Anita Rossovich, to begin taking pictures with David and even to do some assignments on her own. Her skill and feeling for children were evident in Anita's work, and this became very clear to Burton when one parent came to pick up her pictures. "I was giving a mother a photograph she'd ordered, and I remember telling her how I was really pleased at the way I'd captured her child," he recalls.

"After she left, Anita came to me and said, 'I took that picture'—and I hadn't even known it."

After two-and-a-half years as David's employee, Anita had completed "the process of evolution," as Burton calls it. "All of a sudden I realized that I have a sense of children that I'd shared with Anita, and that she was ready to go out and do it—not under the name of David Burton Photography, but on her own. So I fired her."

Burton quickly adds that he "fired" his chief assistant in order to launch her in an independent career—and because he and his wife, Kay, a college professor, had bought an old creamery, and he planned to take a year off to remodel this new home. Consequently, Rossovich lost her job and gained most of David's clients, plus David's promise that if she needed help, he'd work for her on an hourly basis.

"When I hired two employees—the other one quit to go to law school—I certainly didn't plan to leave my business with one of them," says David. "But I'm really pleased that I got somebody else going."

Ueltzen's home-based business began when the phone company asked him to do a short-term job on contract and then asked if he could suggest other people for similar work. He put together a group of experienced professionals and now supplies expert workers for telephone company projects.

Source books published by community groups also identify a broad range of services and skills that are available independently. Some of these reference sources focus on specific types of work, such as the directories of graphic artists and photographers published in many cities. Others have a wider scope, such as the *People's Yellow Pages* or the *Women's Yellow Pages.* The more traditional reference guides, such as the telephone book's Yellow Pages and local newspapers, also are sources of assistance.

If a workstead businessperson plans to hire employees, he can use the same networks to inquire about experienced available help. Employment agencies can supply able workers, and they screen applicants for the employer. Commercial agencies often charge employers a fee, but state and local public agencies do not. Some communities fund programs to train workers for specific jobs; for example, Mike Young hired several key punch operators for his computer mailing list business from a local employment development program. Many communities have programs to place high school and college students in part-time jobs, and other public agencies match qualified handicapped persons with employers' needs.

Sometimes a good employee comes as a byproduct of another activity. Publicist Candice Jacobson, for instance, recently taught an evening adult-education class in the basics of publicity and media contacts. This led not only to more work for Jacobson—from the added exposure the teaching gave her—but also to an assistant, who was the best pupil in the class. Guitar maker Charles Fox had a similar experience when he decided to concentrate on research in guitar design and cut down on his teaching. Now Fox does research and administers his year-round series of guitar-making workshops, while his former apprentice is the salaried instructor for Fox's workstead school.

Communications and Support Systems

Staying in touch and staying in operation are two workstead essentials that require special attention. Although there may be a limited variety of services available in residential areas or other unusual business locations, a resourceful worksteader can find adequate support systems for her business.

Telephone

The phone is a key element in most worksteads. It compensates for distance and provides quick, personal contact. Many worksteaders install more than one telephone line, utilizing the push-button system that holds calls. Some families also maintain a private line—usually a separate number and phone—that is not used for business. This way, they can receive personal calls during and after business hours and have the business phone covered by an answering service or machine. Multiple-line telephones usually require a trunk line, which may have to be installed at the worksteader's expense. In some residential areas there is a less costly service that signals a person who is using his phone that a second call is coming in; the person can't put the first call on hold, but he can tell the person he's talking to that he'll call right back and then hang up and answer the second call. Worksteaders in rural areas have fewer options in phone service, but those beyond the reach of standard lines can use radio-telephones, ham radios, or even CBs.

Covering the phone is also essential for any business. Particularly for one-person worksteads, an answering service or machine permits flexibility without the loss of potential clients. A ringing telephone is both distracting and tempting—few of us can resist curiosity and just let it ring. But a worksteader also must have time to concentrate, without interruptions, which an answering machine or answering service makes possible.

Businesspeople have varying opinions about which type of phone coverage is better—the human contact of an answering service or the recorded greeting and message tape. The recorder isn't likely to garble messages, but some people dislike the process of speaking to a machine and will hang up rather than leave a message. One advantage of an answering service is that the worksteader can unplug his telephone, if it's the relatively new modular type, and be spared the ringing, while the service still answers his calls. An answering service provides a human touch, but with it comes the potential for human fallibility in answering the phone after the caller has hung up and in overlooking messages or transcribing them incorrectly. The answering machine is impersonal and slightly traumatic for some callers, but it offers round-the-clock coverage that is otherwise quite costly. The choice is largely one of personal preference, although if the business phone requires such coverage for more than a year, an answering machine is the more economical alternative: a service costs from $20 to $35 per month; a machine costs from $150 to $350, depending on its features.

For businesses that require a great deal of long-distance phoning, there are several reduced-rate alternatives. The Bell System offers intrastate tie lines and interstate WATS lines for flat fees (a minimum of $200 per month, plus $15 to $20 per hour after the first ten hours of use). Some of Ma Bell's competitors offer lower-cost service to many areas; ITT charges a basic rate of $10 per month and about 30¢ to 40¢ per minute according to distance, while Southern Pacific charges $25 per month plus 8¢ to 30¢ per minute. Most of these private telecommunications networks require touch-tone phones, which are supplied by the phone company for a one-time installation fee. There also is a firm with a toll-free 800 number—Toll Free American in Ft. Lauderdale, Florida—that serves as a nationwide answering service and mails the messages to its clients each day.

Mail and Delivery Services

In spite of all our complaints about it, the U.S. Mail keeps most businesses functioning and most bill-payers out of trouble. Workstead businesses are especially reliant on the mail, even if their out-of-the-way locations mean that service is less comprehensive than it is in commercial areas. One alternative to once-a-day home delivery is rental of a post office box; mail is sorted and placed in boxes several times each day, an advantage for high-volume, rush-order businesses. The trip to a post office box can be a welcome break in routine, too, but if the only available boxes are in a cross-town post office, the mail run could become a commute. Anita Scott of San Francisco Book Company must drive several miles to pick up her firm's mail, but usually she combines this trip with taking her children to school or doing errands.

A postage scale is a necessity for workstead businesses with any volume of outgoing mail, and a time-saving postage meter should be considered. A worksteader who regularly mails parcels of various sizes can simplify his relations with the post office by talking with a local postmaster and getting clear instructions about types of mail service and sizes and weights of packages.

Private delivery firms often are competitive with postal rates and usually offer door-to-door service. United Parcel Service (UPS), for example, covers most parts of the United States and Canada, and its trucks deliver to the residential sections of urban and suburban areas. For a fee of about two dollars per week, UPS also will pick up parcels in residential areas that are near the drivers' established routes; the worksteader puts a "UPS" sign in a window or door so that the driver knows when a shipment is ready. In rural locations, UPS often establishes a pickup and delivery point at a business such as a hardware or agricultural supply store.

Michael and Justine Toms, whose three-story workstead building houses the offices and studio of their New Dimensions Radio network, have hired a private firm that picks up mail from their post office box across town and delivers it to their workstead every day. "The mail service has been a tremendous time-saver for our business," Justine Toms notes, "and it saves money because one of us doesn't have to take the time to drive to the post office and be away from work."

Other delivery firms offer overnight air service and door-to-door delivery of small and large packages. For parcels that require speedy delivery and careful handling, these services are surprisingly inexpensive—a one-pound package can go coast-to-coast for about fifteen dollars. Some air express companies, such as Federal Express

Sandy Kaufman:

Closet Communications

Fast and accurate global communications are essential to Sandy Kaufman, who is both an importer and exporter as well as a consultant to overseas firms. Although he travels to Japan, Europe, and South America on occasion, most of Kaufman's globe-trotting is by wire—from the Telex machine in a closet of his Manhattan loft. "This Telex is like an umbilical cord in our business," Sandy notes. "It's almost the same thing as a telephone, except you're reading it."

The printer that reproduces the Telex messages is noisy, which accounts for the machine's location. If the Telex were placed in the office end of the family's 2500-square-foot loft, it would require cushioning with an acoustical shield so that phone calls and normal conversations could be heard. And because Japan's business day is

our night, many messages arrive between midnight and eight in the morning, which makes the machine's cloistered setting necessary for the family's sleep. If Sandy is expecting an urgent wire, however, he can set an alarm on the machine that will alert him when a message comes in from that source.

The Telex message also is a written confirmation of business transactions, which makes it preferable to telephone calls in most instances. "A Telex is registered as to date and time, and there's a computer record of the registry numbers; so it can be made a legal contract," Kaufman states.

Sandy points out that he uses a kind of shorthand in these messages—numbering each item and question so that his correspondent can reply easily and succinctly. The Telex system, which Kaufman rents from RCA, is charged on a message-unit basis, so efficiency is also economical in these communications.

Telex communication fits the workstead style well, Sandy Kaufman observes, because it keeps him in touch with business with a minimum of intrusion on his leisure time: "In a way, our legal contracts are made at three o'clock in the morning in our closet while we're sleeping."

and Emery Air Freight, also supply envelopes and shipping boxes. In addition, there are local delivery services that operate during week-day business hours in most communities and twenty-four-hour messenger services that are available in some metropolitan areas.

Electronic Aids

In *The Third Wave,* a sweeping analysis of our collective future, author Alvin Toffler cites "the electronic cottage" as an important work place in coming years. Toffler predicts that the costs of "tele-commuting"—moving information via electronic networks, including satellites—soon will be lower than transportation costs to move workers into centralized work areas. Already, he notes, the energy savings are enormous: one Los Angeles study revealed that a commuter alone in his car uses twenty-nine times more energy to get to work than an electronic information system would consume in getting the work to him.

The devices needed to move information have been developed, but most are not yet within the economic reach of worksteaders. The process of integrating home electronics has begun, however, and often with the help of large corporations that can afford the hardware. Buck Kales, a computer consultant, built his home computer but purchased a used companion printer from the Tymshare company. A number of stockbrokers have been able to buy or borrow ticker-tape machines for their workstead offices, and Kales and many home-based computer professionals have electronic access to large data banks elsewhere. In addition, a growing number of large firms, including Chicago's Continental Bank, are experimenting with home work programs for information-processing employees.

Supplies

In addition to the professional tools and materials for any business, every worksteader needs routine office or shop supplies for her operation. Location may prevent a bike ride or stroll to the supply center, but many such firms will take phone or mail orders and deliver the merchandise (usually with a minimum order of, say, twenty-five dollars). If the worksteader manufactures products for retail sales, she may purchase materials from wholesale supply centers, which also usually have a minimum purchase amount. There also are a number of mail-order office and industrial supply firms that publish catalogs and are competitively priced. Ordering from these firms can mean significant savings, but requires advance planning to allow for shipping time. When supplies are needed urgently and the worksteader can't spare the time to go get them, a phone order and a local messenger service can be the solution.

Services

Most repair and maintenance services are available to worksteads, whatever their location. Servicepeople for major brands of office

Bob and Edith Hand:
The Neighbors Bring Birds

Bob and Edith Hand are full-time wood-carvers, who fashion uncannily lifelike shore and game birds in their home workshop. They work from sketches and photos of birds they've seen or, preferably, from a frozen or stuffed bird. The Hands' Sag Harbor, New York, location—near the tip of Long Island—is an advantage in this respect, for it's a stopover for migrating waterfowl and the year-round home of many species of birds.

The professional couple (and their two sons, who are novice carvers) compete in several major shows each year, and the judging in these events is rigorous. That's why they need the real birds to work from, Edith explains: "Pictures aren't going to show you the bird itself—they don't show the back, or underneath it. You really need the stuffed bird or a frozen bird, so you can actually pick up the feathers and look and almost dissect it sometimes."

The Hands acquire these frozen or stuffed birds largely from neighbors and friends who know what their vocation requires. "One duck we just got was caught in a muskrat trap and drowned," Bob notes. "The trapper brought it right over here, and we froze it so the bird wouldn't lose its color." Edith adds: "People bring different specimens to us to carve from or to do our research. They call us up all the time and ask if we need this one or that one. So they keep us with a steady supply of research material."

machines usually will make "house calls" within a few days of a phone request. Occasionally, the person who dispatches repair technicians may be reluctant to send someone because of a business's residential location, but the worksteader can be prepared with the names of nearby institutions, such as hospitals or banks, that have the same type of equipment requiring the same service. If the dispatcher still demurs, the worksteader can call or drop in at the neighboring hospital or bank and get the name of the appropriate serviceperson, then call that person for an appointment.

More specialized services can be arranged for worksteads, too. For example, Dr. Toni Novick has limited lab facilities in her home medical office, so she has arranged for these services at a nearby laboratory that also serves a community medical center. The lab's messenger comes to her office each weekday to pick up samples and deliver test results.

Location

Because most worksteads begin as fledgling operations with limited budgets, their locations are wherever the occupants happen to live. So the location of a workstead is seldom a matter of selection or prior planning, but rather an exercise in making do. And, surprisingly often, this pays off—in a "homemade" image for the business, a welcome change of setting for clients, and a work place without distractions.

If a worksteader can choose his location—when selecting a site for a new workstead or when moving his business to a larger place, for example—his work experience will suggest many features to look for in the new quarters. Though he probably will concentrate his attention on the potential work areas of a new home, apartment, or loft, he also should consider the location as a place to live—in fact, to spend twenty-four hours a day. Other family members' needs, such as access to schools and public transportation, also may influence the choice of a workstead location. Often, a large, empty space with "great possibilities" seems tempting, but the worksteader should realistically assess his ability to keep the business going and simultaneously create a comfortable home in such a setting.

Wherever it is, a workstead's location can be used as an asset. The fact that the place of business is also a home can attract clients, and a residential location can be a real convenience for customers in nearby areas. Off-the-beaten-track locations can be presented in a positive light, with clever maps that show local landmarks, schedules of public transportation serving the area, and even a bit of history about the building or neighborhood. Island Press, a workstead book publisher, calls attention to its location in the obscure northern California town of Covelo by including some historical facts about the area in each of its catalogs. This technique can function as highly effective advertising, because a customer will remember the odd facts about a location and thus also remember the business that's there.

Ruth Pearson:

Location by Inspiration

Ruth Pearson had taught dance to children for twenty-five years when she decided to open a full-time day school, and she knew exactly where it should be located: "When the library was built, I thought, 'There ought to be a school near here, because it's a beautiful library, and children ought to know that places of learning can be beautiful.' All the libraries in the schools look like jails—and environment is everything."

Locating her school, called Inverness Day School, next to the Carmichael, California, library was no easy feat. Ruth and her husband, Stan, had long wanted the lot there, which had a very small, very dilapidated house on it. When the owner of that property died, the Pearsons began a series of complicated, but ultimately successful, negotiations to buy it from the three heirs. Then they found a house for sale a few miles away; the twenty-year-old cottage had an amazing price (even in 1973) of $3000, but carried a stipulation that the buyers had to move it from its present location. "We looked at this house," Ruth recalls, "and Stan said, 'You know, this has a screened porch. We could knock out this wall and extend the porch out and have a big art room and a room for dancing.' So that's what we did."

Ruth and Stan had the old house on their new property torn down and the cottage moved there, only to have the "school" sit on its moving jacks during three months of struggling for bureaucratic approvals. Fortunately, the Pearsons had not yet sold their family home nearby, so they had a place to live in this interim. After the school opened, the couple sold their family home and moved into the vacant bedroom of the schoolhouse. "We had a bed and a soft chair, a desk and a bureau, and about five lamps, so the room seemed bigger with all the light," Ruth notes. "And we 'stored' our art by hanging it all over the walls of the bedroom and the school office. We even had a piece of sculpture in the bathroom."

After three years of "accommodating," as Ruth calls it—which included Stan's building extra legs for the kids' tables so that they could be raised to adult height for the couple's dining or entertaining—the Pearsons added living quarters at one end of the school. Now they share a living room, small kitchen, patio, and bedroom in a compact space that's dominated by a large stone fireplace. "It's like a honeymoon cottage," Ruth says. "My husband and I are living as if we're twenty-four years old."

Ruth Pearson shares her youthful outlook with seven part-time teachers and thirty preschool, kindergarten, and first-grade children in the school's two half-day sessions. She still teaches dance—in the open room of the school or, for large groups, in the church located on the other side of the school from the library. "I love to teach, but I don't like loading my car every day and driving out somewhere," Ruth states. "I want to do it right here. After all these years of teaching, the classes have come to me."

Space

Every worksteader's challenge is to make the best use of available space. The variations are almost infinite, but most worksteaders face one of two general situations: an unused corner of conventional residential space must be transformed into an office, or a large open space must be divided into distinct working and living areas. In both of these situations, the accompanying problems usually include where to store all the business paraphernalia and how to find a quiet spot where work doesn't intrude.

Work Areas

Having designed several of his own offices at home, architect John Edwards is prepared by experience and profession to suggest possibilities for workstead offices. In "Space Planning for a Home Office" (see pages 104–106), Edwards has provided drawings and accompanying explanations for three basic space arrangements: a closet office, work space in part of a room, and a spare room converted to an office. The "bunker office" (part of a room) is particularly adaptable to loft or warehouse spaces as well.

Storage

Once a work area has been established, the next challenge often is to locate adequate and accessible storage space. In some instances the office's physical boundaries can double as storage, such as bunker partitions composed of bookcases or filing cabinets. Other worksteaders accumulate much more material and inactive files than an office area can accommodate and must expand their storage (and tax-deductible square footage) to include a garage, basement, or attic.

Some worksteaders combine storage with living areas. For example, writer-publisher Malcolm Margolin's home also functions as his warehouse for new books. When a shipment is delivered, the boxes are stacked ceiling-high in the dining room, as well as in closets and elsewhere. The dining-room book supply goes first, and this provides a sort of family participation. "Everybody shares a feeling of accomplishment when the boxes disappear," Malcolm states. "There's a visceral sense that things are going well." Margolin also utilizes books as furniture; each family member's bed is supported by several full cartons. In addition, the unused bathtub next to his office is filled with old correspondence and current promotional materials.

Advertising executive Leon Henry, who began his business at home and formerly published a newsletter for home-based businesspeople, suggests that a basic part of any storage system is the wastebasket. This resource, when used often enough, adds to the neatness

Toni Novick:

The Doctor Is In

Toni Novick, M.D., has practiced gynecology and family planning in her home office for twenty years. At the time she and her husband bought their house in suburban New Rochelle, New York, the contractor had not yet completed the building, so they were able to have one part of the garage finished as a medical office.

"We weren't making any great structural alterations in the house," Dr. Novick recalls, "because we couldn't afford that." The would-have-been garage was divided into three small rooms and a central open area: there's a reception room, a treatment room, a bathroom (which adjoins the house), and the central hub where Toni's desk is located. As she notes, this open area is "the secretarial office, the account-

ing, the laboratory, the sterilizing, and the telephone room."

This basic arrangement has served the doctor well, although she has made a few changes. Dr. Novick added a closet that serves as a room divider and provides storage for supplies, files, and reference books. "One of the things I enjoy is making things very efficient," Toni states. "When I get the space to do its best, then the woman who works for me says, 'You have a way of putting things in the closet—everything's in there, but you can see everything.' That helps her, too."

Efficiency is necessary in such a compact space, and so is privacy. One early problem that Toni discovered in her office arrangement was that sound carried through the

wall between her consultation area (the desk) and the reception room, so discussions with patients could be overheard by people waiting. She solved this initially by putting a radio in the waiting room. "But the people in the waiting room would turn it off, so then we put the radio in my office and installed speakers in the waiting room," she reports. This solution led to another problem, however—the sound from the radio carried into the consultation area, so Toni and Larry Novick added a cork wall on the office side to soften the sounds and add to the privacy of patient-doctor discussions.

The Novicks plan to build a new home now that their four children are grown, and Toni says that they will double the size of her office. She also points out that one thing they'll include in the next office is heavy-duty wiring. "When we had the builder make this office, he asked if we wanted more wiring, and we said no, because I didn't have any X-ray equipment or other big units then. But I did get some machines, and when I used to have evening hours, if I'd use one of my instruments it would affect the television and sometimes make the lights go off in the house."

Dr. Novick's home practice has affected the family in other ways, too. She recalls that there have been many times when a patient would call during supper. "The children would know when I was talking on the office phone to hold down the conversation, and they wouldn't mind hearing different discussions concerning discharges, blood, fever—all those things were being said at the same time they were sitting there eating dinner."

Despite the family adjustments and limited office space, Toni Novick wouldn't care to work elsewhere. "I love having my own business at home, even though I get to do the junk as well as the hot-shot stuff. I enjoy being my own boss. I even encourage my children to try to have that lifestyle, to be their own boss."

of a workstead office and helps keep any storage and filing system from bursting at the seams.

Although workstead storage needs are highly individual, depending on a worksteader's location and profession, some of the widely used shelf and cabinet systems can be adapted to most situations. Two books that highlight a variety of ideas and do-it-yourself projects are both titled, logically enough, *Storage.* One is written by Melinda Davis and published by House and Garden; the other is compiled by the editors of *Sunset* magazine and published by Lane Publishing. Many contemporary furniture and design stores also feature modular storage systems, including some that are easily assembled at home. Conran's, a New York store, publishes a mail-order catalog that contains a large selection of storage units and modular systems (see "Resources" for address).

Privacy

An emphasis on work areas and storage systems should not overshadow an equally important part of any workstead's space—its private areas. Even people who enjoy living close to their work need freedom from it at times, and, ideally, workers and family members should not have to leave home for leisure moments. Thus, every workstead should include a comfortable area that is out of sight and earshot of the work space.

Homes and apartments incorporate such private spaces most easily, because they are divided into relatively small, insulated rooms. In large studios or lofts, private areas can be created by installing doors or curtains or walls. Acoustics are especially important in converted industrial or commercial buildings, where sound carries easily and reverberates because of hard surfaces and high ceilings. Soft furnishings, rugs, and heavy draperies (as room dividers, if appropriate) can help absorb sound and increase the livability as well as the privacy of these spaces.

Privacy is particularly important within workstead families. The working and nonworking family members must respect each other's privacy, which can be facilitated by a practical separation of living and work areas.

In group living situations, a workstead business may involve some residents, while others who live in the building have jobs elsewhere or work at different home occupations. On occasion, these differing uses of the shared space can conflict, with the result that one workstead business disrupts the private living space and comfort of the people who aren't part of that business. Such problems require straightforward discussion among all concerned, but often a strict delineation of "living-only" areas can resolve the conflict. For example, the living room and one bathroom might be off limits to workstead employees, customers, or group meetings, so that anyone using these rooms can be assured the privacy of a space that is free from the intrusion of someone else's work.

Eleanore Ramsey:

Doors Make Good Neighbors

Hand bookbinder Eleanore Ramsey shares a six-room and two-bath apartment with her own bindery, another binder's studio, and he students who come there for lessons. One compromise Eleanore has made is sleeping on a Japanese futon in her living room; during the day she conceals the futon and bedding behind a sofa. "I'd really like to have a bedroom," she says, "but we only have so much room, and I have to share the bindery in order to pay the rent."

The former bedroom was converted to a studio, where Ramsey designs and produces one-of-a-kind works in the French tradition of hand binding. Her collection of unusual-looking tools and weights covers most of the walls and work surfaces of the room, although Eleanore does store her clothes in the studio's closet. Except for any project in progress, all of her books are stored in boxes, away from the dust and sunlight that could damage them.

The three rooms beyond Ramsey's studio are her bindery-mate's work area and the teaching rooms, which both binders use for instruction. To ensure privacy and quiet for her own workstead quarters, Eleanore has installed a door in the hallway between her studio and the teaching rooms. "The students have keys to the apartment so that they can work here whenever they have time," she explains. "Sometimes they come when I want to sleep or just want time to myself, and then I close the door—and the students know not to disturb me."

When the hall door is open, students often come to Eleanore with questions, and the binder who shares her space also joins her to share ideas and company. Eleanore appreciates the balance she has created between privacy and collaboration, because her work and temperament tend to make her reclusive. "It's just that I get carried away with my work," Eleanore observes. "It'd be really easy for me to go six months without coming out of the house. So it's nice to hear another voice."

Space Planning for a Home Office
by John Edwards

Home office space generally consists of whatever is available. The amount of space required is as varied as the task or profession involved, and the space usually is dictated by the dimensions of paper or material used in one's work. (Drafting or sewing requires a large desk area, medical examination requires an exam table, and so forth.) Needs for storage, work, and file areas must be determined. All businesses, however, involve standard-size (8 ½″ by 11″) paper of some kind—for letters, billings, and records—as well as storage of reference books and materials.

Work areas and storage. The most commonly available spaces in most homes are closets, extra space within a large room, and spare bedrooms. The most universal office elements are desktops and bookshelves. Bookshelves may be expanded further to include cabinets (shelves with doors), files, and drawers (sliding shelves) —storage spaces that permit you to stack materials in an orderly manner.

Ways to make a small space larger. Bright white paint provides more reflected light than darker colors and thus makes a room seem larger. Or paint one wall a different color, use mirrors, or install spotlights to highlight areas of interest in the room. Art panels can be

Closet Office

made to look like windows if you have a room without a source of natural light.

Lighting. Proper lighting is a high priority for any home office. If you use fluorescent lights, purchase "full spectrum" bulbs; they are more like natural light. Switching off an office light also can turn a mental switch, signifying that work is over, while the intensity of other "living" lights grows.

Entryways. The entryway to your office provides the first impression the client has as he or she comes to your home business. One solution is to make use of graphics—a framed panel or roll-up shade—to display your professional logo or business name. After hours the panel or shade can be flipped over or rolled up to transform the office back into a home. These graphic devices also can serve to screen off living areas of the home and direct the client toward your office area.

Closet office. With a limited amount of space available and required, a closet can do the job. When your work is done, you can close the door. The most practical type of door is a folding door because of its small size when opened. If the closet has sliding or swinging doors, you might consider replacing them with folding doors or with a roll-up curtain of some type.

Most closets are at least two feet deep, which is wide enough for a desktop. A desktop can be as simple as a door placed on sawhorses

Bunker Office

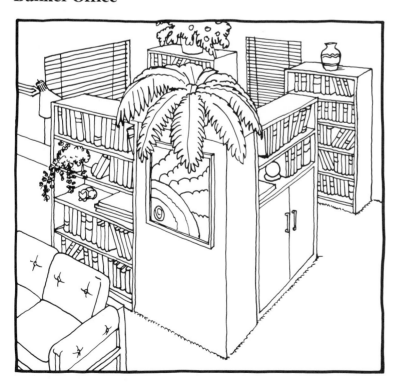

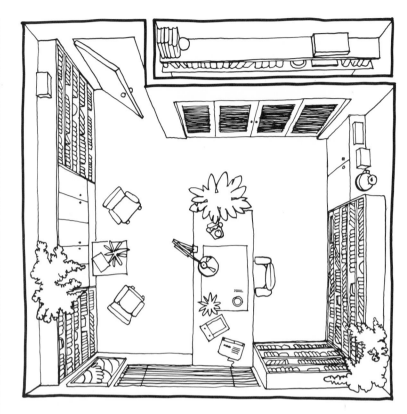

Spare-Room Office

or, better yet, on a stack of drawers, shelves, or cabinets. Shelves and bookcases can be made inexpensively from particle board or plywood and painted, or you can use "high-tech" industrial shelf units. Other sources of shelves are garage sales and flea markets. If the shelves aren't quite the color or shape to match your decor, paint them all the same color and they will blend together. They will lose their distracting and clashing qualities yet maintain interesting shapes and forms.

Bunker office. The "room within a room" concept is practical if you have one large room that you can afford to subdivide. In this configuration, bookcases themselves become walls. By creating a "bunker," you can leave work behind the ramparts, while the bookcases serve the space outside and the backs of bookcases become spaces to hang paintings and bulletin boards. The room divider can be as simple as a single bookcase wall placed behind a sofa or bed. Plants and lighting also can be used to help separate the living and work areas.

Spare-room office. A spare room provides ideal office space and combines the features of both the closet and bunker setups. Closet space can be used for cluttered paperwork and to store materials, with the entire room serving as your bunker. Once you turn off the lights and close the door, you're home!

John Edwards:

Informality at Work

For the past four years, architect John Edwards has done design and consulting work from the small house he shares with his wife, Rita Cahill. Edwards's workstead space reflects the order and openness that are characteristic of his profession: he has fashioned a three-sided "bunker" office of shelves, drawing board, and light table. A swivel chair puts him within easy reach of these work surfaces, as well as the neat rows of templates and tools above his drawing board. Drawings for current projects are mounted on one wall of the office; other drawings are stored out of sight in a hall cupboard.

The living room is easily visible from John's work area, and its furnishings provide a sort of visual coffee break for him. This momentary change of scene is often necessary, Edwards notes, because his work requires intense concentration. "I get spaced out after three hours of solving simultaneous equations," the architect says, "so I need to break it up, go in the other room, or have lunch with a friend."

John finds the workstead arrangement well suited to his personal style. He prefers being at home to a more formal commercial environment, because "you don't have to present a big public image—you don't have to be out there shaking hands and smiling every day." A home office also permits him to dress informally, giving him the opportunity to wear the tropical shirts he loves to collect. "When you think about it," John muses, "it's ridiculous to get up and take a shower and put on a suit to walk into the next room."

Loft living/loft working. Import-export consultant Sandy Kaufman and his family have easily fit a three-person office with limited storage and spacious living quarters in their New York loft. Kaufman particularly enjoys the flexibility of his open living area: "If we don't like the living room where it is, we just move it." The Kaufmans chose to keep their space open to take advantage of city views at both ends of the loft; instead of additional walls, Sandy installed a movable fiberboard partition to screen off his work space on weekdays. Creating bedrooms posed another challenge in this former industrial space. To supply light and air to son Michael's sleeping area, which has no windows, Kaufman cut a large circular hole in the foot-thick wall that opens on the living room.

Raising the roof beams. With her brother's and sister's families, weaver Alexandra Jacopetti bought an old summer camp, where she lives and works with her extended family. Alexandra chose the former infirmary building for her studio, where the thirteen-by-twenty-foot high-warp loom she uses could be built. Installation of the loom required raising the infirmary's roof and replacing the building's center beam with the top beam of the loom. This structural alteration resulted in a work area that allows Jacopetti to make large weavings and tapestries for such monumental spaces as the Pennzoil Towers in Houston.

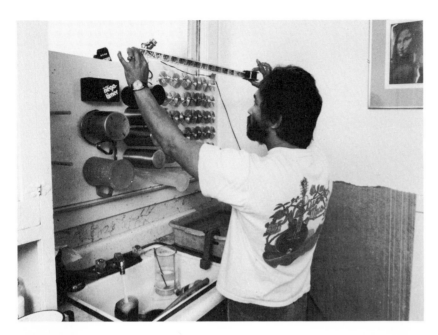

Portable studio. Photographer Emilio Mercado has devoted his three-room apartment to his career. He utilizes the large main room as a studio, keeping it free of everything but a few props for portfolio or advertising photos; the other large room provides storage and display space for his work, while the closet and kitchen are filled with darkroom equipment and chemicals. All of Mercado's darkroom setup, storage shelves, and even his sleeping futon are portable, in keeping with his strong sense of staying free to travel or take on photography assignments. "I think it's a hindrance to have lots of furniture and things," Emilio states, "because it doesn't do me any good in my work. The important thing is to be doing the work."

Long-distance efficiency. Barbara Dean's workstead is a canvas-sided yurt that she built ten years ago on the 670-acre ranch she shares with several other owners. From this compact circular space Barbara runs a small publishing firm, Island Press, without benefit of electricity or telephone. All of Dean's routine business is handled by mail—the ranch has a daily delivery—and her once-a-week trips to the nearest town (twenty-five miles on a dirt road) allow for phone calls to authors, suppliers, and the three part-time staff members who live about a six-hour drive from the yurt headquarters. Dean finds this arrangement workable and compatible with her desire for a life close to nature: "I'm able to live where I'm happiest, and working in the yurt forces me to be organized as well. I think most business dealings are more efficient by mail, because everything is in writing, and the correspondents have to think through what they are putting down on paper."

White and light. Eleanor Johnson and Judah Kataloni are directors of the experimental theatre Emmatroupe, which rehearses and performs in their Greenwich Village loft. This former hat factory encompasses one floor of two narrow buildings, which are separated by a brick wall with a fire door at one end. On one side of the wall is the open theatre area, and on the other side are Judah's and Eleanor's offices and living quarters; the wall and door meet fire-safety regulations and provide ample privacy for the activities in each space. To build privacy into the living area–office side of the loft without impeding air circulation, Kataloni designed and installed six-foot-high wooden partitions, which have been painted white to retain a feeling of light and openness.

Studio under the roof. Ken Nordine installed a sophisticated recording studio in the attic of his rambling Chicago home so that he could do all of his writing, producing, and recording there. Nordine, whose pleasing bass voice is familiar to anyone who hears radio or television commercials, creates "sound collages" in his home studio, combining music, words, and a wide array of vocal sound effects into finished pieces for commercials, a radio program, and record albums. When he made the change to a workstead, installation of the several tons of recording equipment presented a major problem, because most of the machinery was too big to be carried three floors up a winding staircase. Nordine solved this problem by reinforcing the attic floor and extending one of the steel beams several feet outside the house; then all the heavy equipment was hauled up to the attic by block-and-tackle.

Making every inch count. Financial consultant E. B. Cochran is a prolific writer and researcher, to which his extensive files are testimony. In addition to the 2000-square-foot office area and the five large filing cabinets in his secretary's office, Cochran has expanded into the basement's furnace room and one wall of his garage for storage of more papers and reference materials.

Let there be dark. Filmmaker Bob Charlton converted the ground floor of his workstead into a movie theatre. In addition to building a wall of plywood-backed shelves to separate the projection booth from the viewing area, Charlton had to cover a row of picture windows along one side of his hillside home. Rather than buying expensive, custom-made "blackout" shades or draperies, the filmmaker constructed several floor-to-ceiling plywood panels to cover the windows. These panels can be opened in sections or removed without damage to the walls.

Storefront storage. One drama group solved the problem of limited storage space in a storefront theatre by hanging their props and costumes on one wall. This "open storage" decorates the home of Iris Landsberg and Adele Prandini, directors of the feminist theatre company It's Just a Stage. The original storage areas for this workstead—two small rooms at the back of the theatre—have been converted into a kitchen-office and a loft-bedroom for the residents. Because both women teach classes and the company rehearses and performs on the open floor area, none of this precious space could be appropriated for storage.

Legalities

Apart from taxes (see "Finances"), most of the legalities a work-steader must contend with—including zoning regulations, building and safety codes, and the purchase of business licenses and insurance—occur at the local level. Whether a worksteader resides in a city, town, or unincorporated area, the appropriate regulations will be applied to his business, according to what the business is, where it is located, and how it affects the people around it.

Because worksteads concentrate business and residence in one place, they have a dual relationship with local government—which at times can double the frustration of dealing with a bureaucracy. Of course, some worksteaders choose not to make their businesses obvious to local governing bodies, but they run the risk of eventual discovery and, perhaps, penalty. For most people, a relationship with governmental bureaucracy is inevitable, and worksteaders will be most successful in this relationship if they take responsibility for knowing the regulations and rights that apply to running their businesses.

Zoning

Zoning laws had their inspiration in disaster. Amsterdam instituted the first formal zoning plan after a fire that destroyed much of the city in 1451; the new regulations placed artisans' and middle-class living areas close together, centralized businesses, and set off the homes of wealthy residents in a separate area. In the United States, zoning became prominent in the early twentieth century. A zoning code established by New York City in 1916 set a precedent that the rest of the nation followed. This code was based more on social and aesthetic concerns than on safety—the owners of stately commercial and residential buildings on Fifth Avenue didn't want industrial buildings invading their territory. The idea spread quickly; within a decade 591 American cities had adapted this new code to their needs.

Zoning Categories

Zoning essentially spells out what activities are permitted in specific portions of a city, county, or other legally defined area. Most such areas are divided into three general zoning categories—residential, commercial, and industrial—which often are subdivided to make further distinctions, such as light or heavy industrial use and single-family or multiple-unit residences.

Whatever the zoning category, there's likely to be a workstead within its boundaries. The requirements for a legal work-live arrangement, however, vary widely from one zone to another and even more widely from one town to the next.

Industrial. Most cities (or towns or counties) rule out any

Carolyn Buck:
Changing the Law

After a zoning-code ruling forced her out of the old barn she had converted to a studio and home, artist Carolyn Buck got angry—angry enough to organize other artists and to get the law changed. Before the 1979 amendment to Berkeley, California, zoning laws that Buck's group sponsored, every workstead required a use permit from the city planning department. The revised law eliminates the permit requirement, except for such group activities as classes, sales, exhibitions, and performances. This change in one local law took more than a year of meetings, constant phone lobbying, and a visible presence at city hall.

Carolyn Buck led these efforts from the attic apartment she has occupied since losing the barn studio. The limited space has halted her painting temporarily, but Buck currently is looking for a building that the artists group can purchase as a co-op. In the meantime, she devotes much of her time to writing poetry and to working for further changes in the legal attitude toward worksteads. For Carolyn, the need is basic: "This whole work-live thing is an opportunity for artists to develop, and if they don't have this opportunity to work without the interruption of constant moving, then it just delays the process of their growth as artists." Buck adds that the combination of living and work space is essential for artists, because "they have an itch, an inner demand to work, and the itch doesn't stop at five o'clock."

residents in industrial sections based on their judgment that the hazards, noise, dirt, and heavyweight traffic of such areas are not conducive to domestic tranquility. In many localities, an exception is made to allow night watchmen to live on the premises of factories and warehouses. That exception has been the legal loophole for many worksteaders whose occupations require the large spaces or heavy equipment found in industrial zones. In fact, live-in "watchmen" who also install a photography studio or a large loom for their work have reclaimed moribund buildings and brought life and commerce to otherwise-abandoned stretches of urban landscape.

Some cities do provide leeway for living in industrial areas, however. Minneapolis sanctions warehouse residences, so long as the fifty feet of ground-floor space nearest the street are used for commercial purposes. Other cities either ignore the illegal occupation of industrial buildings or place a low priority on enforcing these zoning provisions.

Residential. It's easier to conduct business in a residential area than it is to live in an industrial zone. But there are legal "if's" to this situation—if the business can be contained within the home, if it doesn't create excessive traffic or noise, if all the workers are residents of the home in which the business is located. Such restrictions are common in most cities and towns, whether or not the local governing body formally designates home businesses as a zoning subcategory. Other types of regulations applied to residential businesses include: limitation on the size of business space (usually no more than 25 percent of a residence's total square footage); prohibition of a sign or window display, or establishment of a maximum size for such a sign; and prohibition of retail selling and storage of goods within a residence. Access to a home business also is regulated by many zoning ordinances; some cities require that the business office not have a separate entrance, and others specify a separate entryway for clients or service personnel.

Some communities prohibit all types of home business. Often these areas are suburban and well-heeled, where the residents may have voted for such a policy or a strict zoning board has interpreted a residential area as one containing no businesses. Just as often, these exclusively residential suburbs are bordered by towns with more liberal zoning regulations; for example, the New York suburb of Scarsdale frowns on all home businesses, while its neighbor New Rochelle has a number of home-based doctors and dentists.

Similarly, some communities single out certain occupations as permissible or prohibited in a residence. Commonly allowed are clerical, sewing, art, and crafts work, as well as most of the professions—lawyers, doctors, architects, and counselors. Zoning laws also exclude many residential businesses; for example, New York City forbids home electrolysis, stockbrokers, and veterinarians, among others. California has a law to protect entertainment job-seekers from the "casting couch"; although literary agents are permitted to do business from home, no actors' agent or talent scout is

Arthur Monroe:

Using the System

The knit "watch cap" that painter Arthur Monroe wears is partly functional and partly symbolic. It keeps his hair in check but also signifies his official status as night watchman in the 2000-square-foot loft he is converting to a workstead. Monroe and eight others who share this large warehouse building had a close brush with eviction when a building inspector filed a report that claimed the building had too few bathrooms for the number of occupants. In fact, Arthur notes, "We've got nine people here and eleven toilets."

It took considerable time and negotiations to clear the record for these worksteaders. They got help from some city officials, though; a newly elected city council member cut through much of the bureaucratic red tape for them, and a member of the planning department studied the code provisions to find a legal means of allowing the worksteaders to occupy the warehouse. Local arts advocates also supported their efforts, and with the help of a lawyer the group had a successful meeting with planning officials. "One person at that meeting found a clause on the books that allows a night watchman to live on the premises to protect the materials," Monroe recalls. "So they had no difficulty rationalizing our staying here, because we were all individual industrialists and we needed to stay near our materials to guard them."

The experience has led Arthur Monroe to become active in organizing other artists to seek legalization of worksteads in large buildings. This struggle, Monroe asserts, is basic to the artist's place in society: "As artists, until we get that approval, then we're all illegitimate—and that makes us really vulnerable."

permitted to work with clients in a home office.

Perhaps the greatest limitation in virtually all zoning ordinances affecting workstead businesses is the stipulation that no employees from outside the home may work in that residential setting. This regulation often is ignored in small home-based businesses, and in some cases is circumvented by the occasional use of independent contractors rather than employees. The rationale behind this and most other zoning regulations for home businesses is a commitment to preserve the residential character of a neighborhood; thus, a low profile is good business. As one city official advises: "Common sense will avoid many of these problem areas in a home occupation."

Commercial. Communities generally follow a middle path of zoning regulation with regard to commercial areas. Residential buildings are permitted within commercial zones, so the rules for home businesses apply in these settings. The old-fashioned store with living quarters above or behind it usually falls within the legal limits also, because it combines the two activities permitted in the area. Even establishing a residence in a former commercial space is allowed in many cities, though the high rent or price of such property is likely to discourage prospective dwellers.

Exceptions to the Rules

Despite the rather detailed regulations of any community's zoning code, exceptions are always possible. The two common means of exemption from zoning regulations are the use permit and the variance.

A use permit is given by the community's planning or zoning officials after an investigation of the proposed use for a specified property. Often this is an informal process in which a prospective worksteader explains his or her plans and facilities to a planner or committee of officials. In other instances, a use permit may be granted conditionally; for example, Dr. Sheldon Baumrind was given a permit after he provided off-street parking for clients and employees of his orthodontics practice.

A variance usually is needed if a workstead business could have significant impact on a neighborhood. For this reason, extensive review and public hearings are often part of the process. If a dancing teacher wanted to convert his home's garage into a studio for small classes, for instance, he would have to apply for a variance in most areas. Community planning officials would review the application, and a public hearing would be held to sample the opinions of his neighbors regarding traffic, noise, and any other possible disruption of the neighborhood.

Similarly, construction of an addition or separate business building on residential property would require a zoning variance. Accountant Nancy Feiner wanted to expand her home work space by building an addition to her home. She polled the neighbors on her street, and all of them supported her plan. When the accountant

Strategy for Survival:

Landmark Status

Old buildings frequently offer ideal workstead space. Today the forces of inflation and rapidly changing technologies have left hundreds of industrial and commercial buildings unfit for their former uses and unlikely to be converted or upgraded. This is precisely the climate that attracts new occupants who will fit their endeavors to the available space.

As urban demands for space continue to grow and property taxes increase, however, old buildings are in jeopardy. Many have been torn down for new structures or remodeled at great cost to their historic features. But concerned planners, worksteaders, and urban historians have moved to stop the demolition—and, in some cases, even the renovation—of the remaining examples of our architectural heritage.

The landmark strategy can be attempted at the city, state, and national level and has been used by a number of workstead groups. The Goodman Building in San Francisco currently has only partial landmark status; its façade has been so designated, but the residents thus far have been unable to gain similar status for the interior, although the building has been used continuously as an artists residence and work place for almost a century. In Massachusetts, an artists group saved Boston's historic Fenway Studios from the bulldozers by successfully campaigning for its landmark status. Subsequently this group was able to buy the old building with the crucial assistance of two tax breaks: one break, from the state, exempts historic buildings in need of repair from property taxes, decreeing that taxes can be based only on the total rents of a building (rather than its market value); the second break, from the city, postponed these tax payments until the group could arrange financing and a payment schedule for the building's purchase.

Owners of historic structures receive other breaks, too. New York City grants a twelve-year exemption from higher property-tax assessments if the owner rehabilitates the building. And the Massachusetts state building code contains a provision that certain historic structures cannot be expected to meet present code requirements, because they were built long before such standards were established.

Achieving landmark status is usually a lengthy and convoluted process. Each city or state has a variety of requirements and seemingly unending red tape, and the cherished national landmark status can take years to accomplish. Yet the sustained efforts of workstead tenants and their community supporters are always worthwhile, for they ensure preservation of both homes and history.

appeared at the public hearing for her variance petition, however, a neighbor from a nearby street spoke very strongly against her proposal, citing the family character of the area and the traffic that her business might bring. Ordinarily, such opposition would have resulted in a denial of the variance, but Feiner argued that she should not be denied the right to earn a living, and as a single parent with three children at home, she could only be both mother and breadwinner from a home office. The city council agreed with Feiner's point of view and granted a zoning variance.

Mixed-use Zoning

In response to changing economic pressures and growth patterns, many cities are moving toward mixed-use zoning, as exemplified by Minneapolis's law allowing people to live in warehouses so long as at least part of the ground floor is used for commercial enterprise. Such varied use of space dates from Greek and Roman times, when houses commonly contained workshops for artisans or artists. Contemporary students of culture and cities, including Lewis Mumford and Jane Jacobs, have noted that much of a city's vitality comes from the intersection of its divergent activities.

In American cities, the movement to establish the mixed-use concept has been led by artists, primarily because their needs for large spaces, ready access to their work, and low rent have been met by fashioning living quarters in unused industrial buildings. In recent years, mixed-use zoning has been instituted in many places, although in some localities this option has been limited to artists only. In 1979 the California state legislature passed a broad measure that permits cities and counties to adopt special rules for "joint living and working quarters." This legislation does not supersede any local laws, but it clears the way for communities to sanction worksteads. A few months before the state law was passed, San Francisco revised its zoning regulations to permit residency in all sections of the city.

Artists and legislators are not the only proponents of the workstead option. Two realtors in Roswell, Georgia (near Atlanta), have begun development of Barrington Close, a row of eight townhouses that feature commercial space on the ground floor and living quarters above. Florence Bruns and Emily Smith have modeled their one-and-a-half-acre project on an English village of the 1800s, complete with period decor and landscaping. The two partners convinced Roswell's city council to amend zoning laws to permit this mixed use of space, and they have had little trouble finding buyers for the spacious townhouses. Bruns and Smith plan to initiate similar projects when Barrington Close is completed.

As the costs of transportation and business space increase, the workstead option is certain to become more prevalent. It is an alternative that legislators, warehouse dwellers, and real-estate developers have recognized, and the growing prominence of this movement is certain to affect zoning laws as well.

Building and Safety Codes

Although many zoning laws are becoming more flexible, building codes are less likely to change significantly. The intricate regulations concerning building, health, and fire safety in every community represent bureaucracy at its most formidable. These patchworks of law originally were formulated to ensure "fire and life safety," but sometimes the specifications seem designed to support the various building trades as much as to protect dwellers and workers. Further complicating the matter for worksteaders, interpretation of the codes is left to individual inspectors, whose rules may vary with the tenor of their dealings with citizens.

Because of the problems involved in complying with intricate building and safety codes, many worksteaders prefer not to deal with them. The costs of bringing a building "up to code" can be monumental, and some types of work, such as the handling of noxious substances or storage of unstable chemicals, probably would be judged too hazardous to be combined with living quarters in any case. Often, even people in occupations and locations that are technically legal seek to avoid the attention of building department officials, because an inspection could uncover flaws that require correction. Avoiding inspection or enforcement of codes is often a calculated risk, but if real hazards exist, this course is irresponsible in terms of family and community safety.

The specific provisions of building and safety codes vary from one city or county to another, but all of these codes reflect certain basic concerns. Among these are structural safety of a building; construction and exits to meet fire-safety regulations; adequate light and ventilation; minimum electrical and plumbing standards; and, for any residence, heating and kitchen facilities. When an occupation is combined with living quarters, the dwelling must meet the standards for both situations. For example, a painter whose paints and cleaners give off toxic or flammable fumes may be required to install an air-changing fan in her work space and to separate her living area from the work space by a fire-resistant wall.

Industrial and Commercial Buildings

The implications that building and safety codes have for a workstead depend on the original function of the workstead building. Any structure that was not built as a residence initially is likely to require a fair amount of modification to meet the standards for a dwelling. For instance, most codes stipulate that living quarters contain a kitchen, hot water, a heating unit, and minimum plumbing; an old warehouse probably will lack some or all of these amenities, and installing them could require major structural changes.

Large buildings often have another built-in problem: because a group of tenants or owners is needed to pay the high rent or mortgage on these spaces, the stricter building codes that apply to group dwellings usually are enforced. Some of these standards are rigid and impractical, such as the common requirement that each individual living unit include a kitchen and bathroom. Certainly many tenants would prefer sharing a kitchen and even a bathroom to paying for extensive renovations.

Residential Structures

Residential buildings that become worksteads have the expensive dwelling features in place, but exhibit other limitations because they were not originally designed as work places. The owners of a home-based cooking service had to install heavy-duty wiring and commercial kitchen equipment, for example, and the number of people allowed to work or attend cooking demonstrations in this workstead is limited by fire-safety rules. The growth potential of a business is related to the size of a home and the residential-style expansion that can be made there; many worksteaders eventually move their businesses out of a home because expansion is not legally or economically feasible.

Just as living quarters are not permitted in certain industrial settings—because of the materials used in the work—certain occupations are not permitted in residences. Use of flammable chemicals in home work is strictly regulated, usually requiring a special inspection and permit from the fire department. Among these potentially hazardous occupations requiring a fire inspection and permit are woodworking (chiefly because of the dust), candle-making, and auto repair.

Inspection

Building and safety codes are enforced by inspections, which occur at several times during renovation or new construction for a workstead. If no renovation is planned but an inspector rules that certain changes are necessary, the worksteader will have to make the required changes or discontinue his work or residence in the building —whichever is in violation of the codes. Most communities have a system of appeals, however, or, occasionally, a waiver of building-code regulations, and worksteaders have used such appeal processes successfully. As sculptor Tom Gibbs found in his dispute with a Dubuque, Iowa, building inspector, a citizen who has a valid argument can be heard if he is persistent and politely vocal. Gibbs obtained media coverage of his efforts to gain approval of his studio, which he wanted to build on the bedrock of an abandoned quarry near his house, but not adjoining the residence as the law specified. When television and newspaper articles told of the sculptor's difficulties with one building inspector, other citizens came forward to relate similar problems with this bureaucrat. The conclusion of Gibbs's civic struggle was approval of his studio's site and quiet retirement for the inspector.

Artists have long been in the vanguard of the workstead movement. Particularly in European cities, studios and workshops were integrated with living quarters almost routinely; in most places, this was seen as a logical form of construction. Thus, artists did not have to deal with the problem of bringing a work place up to the standards of a dwelling. However, as cities everywhere have become crowded and old buildings replaced by high-rise apartments and sprawling civic structures, these desirable spaces have been disappearing. Even Paris —where, for centuries, artists and their work places have been sacred—bulldozed the heart of its Montparnasse artists community for a new railroad station.

The bulldozers also have struck in many American cities. The urban building boom of the late 1940s and 1950s swallowed up thousands of old structures that had served artists and artisans. When rising costs slowed this tide of urban renewal in the mid-1960s and small manufacturing firms moved to more modern quarters or simply folded, building owners were grateful for the artists who would pay even modest rents for their declining properties.

Invariably the artist-tenants chose to reside illegally in their warehouse or factory work spaces (known collectively as "lofts"), and invariably the landlords looked the other way, avoiding the notice of city inspectors, while the rent checks kept rolling in. When owners discovered that their workstead tenants had made valuable improvements to their spaces, however, these loft dwellings became the objects of controversy. Suddenly these spaces were more desirable as living quarters, and landlords were able to command higher rents than the artist-improvers could pay.

Jim Stratton, a New York artist and loft dweller, gives a detailed history of the artists' struggle for living and work space, in his book *Pioneering in the Urban Wilderness.* Stratton chronicles the Manhattan artists' experience in organizing to achieve legal status for work-live spaces: in the early 1960s, Mayor Robert Wagner issued an executive order creating "artist-in-residence" dwellings, which gave legal sanction to worksteads, but this did not discourage landlords from evicting artist-tenants or demanding ever-higher rents for lofts that artists had improved. So a great number of loft dwellers continued their work-live situation illegally, to avoid confrontations with landlords, building inspectors, and trend-setting realtors.

As loft living became more popular among New York artists, various groups began to fashion alternatives to the renovation-eviction cycle. In the late 1960s, several artists cooperatives were formed; an artist could purchase space and joint ownership of a building, with all the occupants sharing taxes and maintenance costs. The center of this activity was an area bordering Greenwich Village known as "SoHo" for its location south of Houston Street. The artist-residents of SoHo began working as an organization, establishing contacts with city building-department staff members and gaining support in the thriving New York art world.

After many months of meetings and numerous threatened evictions for zoning or building-code violations, the SoHo Artists-Tenants Association won a battle with the bureaucracy in 1971, when artists were legally permitted to occupy lofts in the district. This privilege was reserved for artists only, and potential loft residents had to be certified by a Cultural Affairs panel. In practice, only a small percentage of the area's residents did apply for certification, because the majority wished to maintain their low profile.

The SoHo artists' struggle made news, however, and by the mid-1970s this district of century-old cast-iron-front buildings had become a chic gathering place, with galleries, boutiques, and restaurants. As rents increased everywhere in Manhattan, loft living became more and more attractive, and suddenly some of the artist certifications were being granted to lawyers who also took pictures or social workers who painted a little. Not surprisingly, the rents rose in SoHo, too, and bargain hunters spread into neighboring areas in search of lofts.

Today, urban artists must compete with live-in businesspeople and those who simply want a loft instead of an apartment. High rents have driven many workstead artists from fashionable loft districts, but this vanguard has merely moved on, staking out new territory as usual.

Business Licenses

Every community has some form of business license, and in most places a worksteader must obtain one. If there is a fee for this license, it is modest—generally $15 to $25. Once licensed by a community, businesses are subject to a business tax. In many instances, however, worksteads are exempt from such taxes, either because of their home location or because their volume of receipts falls below the minimum tax level.

In many communities, too, all businesses that have a name other than that of the proprietor must file a "fictitious name statement" with the local government. Usually this involves a small fee, and the worksteader may choose to spend another $50 to $70 for the publication of the business name to protect it from use by another local firm. In most places, however, publication of the name is not required by law.

In practice, many worksteaders do not obtain local business licenses or fictitious name statements and are not noticed by local officials or tax collectors. But if a home-based firm makes news, or begins steady advertising, or otherwise calls public attention to itself, the worksteader undoubtedly will hear from a local official. Usually this results in immediate purchase of a license and possibly some assessment of unpaid business taxes, which are based on gross income.

In addition to standard business licenses, communities often require special licenses for occupations in need of strict regulation. For example, most large cities require special fees, permits, and even fingerprinting of operators of home-based escort services and limousine or taxi services.

Liability and Insurance

Worksteaders engaged in manufacturing should consider insuring their products against injury or damage caused to users. Although such product liability insurance may be costly, it can be purchased for as little as $100 per year for products that are not considered hazardous.

Most other types of insurance coverage for a workstead business can be added as provisions of a standard household policy. Liability coverage for persons connected with a home business can be added as a rider for about $50 per year. Business equipment and inventory also can be protected as part of a household policy, but the worksteader should advise the insurance agent that these items are used in a home business. If the workstead business involves mail-order sales, the firm's mailing list should be insured and a duplicate of it placed in a secure, fireproof location.

An excellent digest of general business-insurance needs is contained in the book *Small-Time Operator*. Author Bernard Kamoroff, an accountant, suggests that business owners consult an independent insurance agent, who can fashion a cost-efficient insurance "package" from among the various companies he or she represents.

Finances _____

By their very nature, most worksteads are small businesses. Thus, the financial needs for these essentially cottage industries are rather modest—at least in comparison to the millions that conglomerates and governments toss around. Yet the need for careful and fruitful use of financial resources is just as great or greater in workstead enterprises, where a misstep could jeopardize current operations and future opportunities for funding.

In fact, worksteaders can benefit from the financial experience and management techniques of the big guys. For example, the thorough planning and evaluation of assets and markets that go into any business's search for capital are especially valuable to a small, home-based venture. In approaching any potential investor, worksteaders should exhibit the same degree of professionalism as larger firms, because they are competing for virtually the same dollars.

Capital

The distinguishing feature of all worksteads—that they combine living and working spaces—can be an asset in the competition for capital. Particularly if his business has been operating for at least six months or a year, a worksteader can demonstrate efficient use of space and resources and clear savings in overhead, which are indicators of good management and a healthy enterprise. A worksteader might even invite potential investors or lenders to his home office, so that they can see how and where he gets the work done. This tactic is an excellent way to convey the attitude that "I have chosen to work from home for good reasons," dispelling any notion that his choice of location was by default.

Sources of Funds

The sources of capital we usually think of first—banks and government programs—are not the most likely supporters of worksteads; several other options should be tried first. No requests for financial backing should be made, however, until the worksteader has developed a detailed business plan (see chart on page 125). In this respect, going through the steps of completing an SBA (Small Business Administration) or bank loan application is useful, for these procedures require a realistic and thorough assessment of the business in question.

Start-up funding is the most difficult form of capital to get—because the business has not yet proved itself—but this is what many worksteaders are seeking. Knowledge of the sources and procedures for obtaining basic capital also is useful to home-based professionals seeking money for established businesses. Therefore, this discussion focuses on the principal ways to raise start-up capital, with these sources presented in the order of the least complicated and time-consuming procedures. Addresses for the organizations mentioned in this section, as well as the titles of several reference works on raising capital, can be found in the "Resources" section of this book. One valuable guide deserves mention here, too—"Financing Small Business," from the Bank of America's *Small Business Reporter* series. This booklet includes a glossary of financial terms, a succinct discussion of the types of financing that can be arranged, and a detailed sample loan "package," which should be studied by anyone seeking business capital.

Personal funds. Many worksteaders are able to launch their businesses with savings, earnings from investments, or family inheritances. Such a nest egg can forestall worry about a means of personal or business survival as a workstead career is becoming established, but it should not lead to complacency. Even though it's her own money, a worksteader would be unwise to squander funds on any business that is not planned and managed prudently.

"Bootstrap" financing. Essentially, bootstrap financing consists of letting the business support itself. This form of capitalizing an enterprise can be nerve-wracking, because it requires watching every penny and stretching credit and resources to their limits, but these measures can keep a business going without loans or outside investors. In some occupations, the customers supply the bootstrap capital, because they make down payments or pay for the product or service before the worksteader's suppliers must be paid. For example, Bob and Edith Hand require an advance deposit of one-half the price of their custom-carved birds, principally to compensate them for the many hours invested in each of these intricate woodcarvings.

Investors. Family and friends are obvious sources of capital for a business, but worksteaders should note that financial dealings often have a way of damaging or at least altering these relationships. Whether the potential investors are acquaintances or strictly business contacts, however, the worksteader should prepare a detailed statement of business plans and budgets and develop a realistic time frame for return of the investment. Quite often an investor will back a business in return for part ownership of it, such as the "silent partner" who shares ownership of Robin Davey's and Bill Welch's Glad Hand Designs.

The Glad Hand investor does not participate in running the business, but other investors may insist on taking a more active role. Worksteaders may face a dilemma in considering such sources of capital; management advice from experienced investors can be very helpful, but an active partner also can change the whole character of a workstead operation. For instance, an especially promising new business may receive backing from one of the many venture capital

Sample Business Plan Outline
for Loan Application

I. Cover Letter
  A. Dollar amount requested
  B. Terms and timing
  C. Type and price of securities
II. Summary
  A. Business description
    1. Name
    2. Location and plant description
    3. Product
    4. Market and competition
    5. Management expertise
  B. Business goals
  C. Summary of financial needs and application of funds
  D. Earnings projections and potential return to investors
III. Market Analysis
  A. Description of total market
  B. Industry trends
  C. Target market
  D. Competition
IV. Products or Services
  A. Description of product line
  B. Proprietary position: patents, copyrights, and legal and technical considerations
  C. Comparison to competitors' products
V. Manufacturing Process (if applicable)
  A. Materials
  B. Sources of supply
C. Production methods
VI. Marketing Strategy
  A. Overall strategy
  B. Pricing policy
  C. Method of selling, distributing, and servicing products
VII. Management Plan
  A. Form of business organization
  B. Board of directors composition
  C. Officers: organization chart and responsibilities
  D. Resumes of key personnel
  E. Staffing plan/number of employees
  F. Facilities plan/planned capital improvements
  G. Operating plan/schedule of upcoming work for next one to two years
VIII. Financial Data
  A. Financial statements (five years to present)
  B. Five-year financial projections (first year by quarters; remaining years annually)
    1. Profit and loss statements
    2. Balance sheets
    3. Cash flow charts
    4. Capital expenditure estimates
  C. Explanation of projections
  D. Key business ratios
  E. Explanation of use and effect of new funds
  F. Potential return to investors; comparison to average return in the industry as a whole

Reprinted with permission from Bank of America NT&SA, "Financing Small Business," *Small Business Reporter,* Vol. 14, No. 10, copyright © 1980.

firms that exist to fund such enterprises and make a profitable return on their investment. Because the venture capital investor's motives may be fast growth and quick return, however, the worksteader eventually might find himself being pushed beyond the boundaries of a comfortable, home-based business.

Banks and other lending institutions. This is where the competition for start-up money becomes especially stiff. Banks are in the lending business, but they are stingy with long-term business loans for unproven enterprises. These institutions base their decisions on a borrower's experience in the proposed business, his personal credit history, and the business's ability to repay. Only a few fledgling worksteads are likely to meet these criteria. A more promising option is a short-term bank loan, which generally is secured by assets such as equity in real estate. This type of loan can provide the seed money that will get a workstead business started.

Savings and loan associations offer mortgages as their primary type of loan, although the federal regulations governing them recently have been relaxed to allow for more latitude in lending. Credit unions offer secured loans to members, though they may not fund businesses as such. Another option is life insurance firms, which will loan money on certain types of policies. The one advantage of such loans is that even the interest can be deferred as long as the premiums are paid. If the loan and interest are not repaid, however, the policy value will be reduced by that amount.

Even if they don't expect approval of the application, workstead businesspeople can still benefit from preparing the required papers for an institutional loan. As Mike Young of Mailing List Services puts it,"One of the cheapest ways of buying an expert opinion is to try to borrow from it. When you borrow from family or friends, you're not given the scrutiny of a man familiar with the risks and opportunities of the marketplace that you get when you borrow from a stranger."

Even in these rather tumultuous economic times, the "strangers" at banks and other lending institutions do approve loans for businesses that have a sound basis, and the prospects seem especially good for funding workstead enterprises that will employ local citizens. Thus, getting a loan may not be impossible, but it requires preparation, persistence, and careful shopping. Interest rates and loan amounts vary from one lender to another, although banks usually require applicants to raise 50 percent of the needed capital on their own.

Government programs. The Small Business Administration is a well-known source of funds for ventures such as worksteads. Like banks, however, the SBA does not automatically approve loans. This agency's definition of "small" includes firms with up to 1500 employees and receipts in the millions, so the competition is fierce. The application process is complex and long, too, taking from three to eight months.

Workstead businesses may qualify for any of several SBA pro-

Mary Anne Holley:

Finding Funding

Business specialist Mary Anne Holley has been participating in one workstead while seeking funding to start one of her own. At present, Mary Anne and her husband, Ed, both experienced businesspeople with MBA degrees, are nearing completion of the financing to begin publication of a new magazine, *BMF,* which will focus on the interests and lifestyles of contemporary blacks. Their magazine venture has been several years in the making, and its backing has been arranged through a Minority Enterprise Small Business Investment Company (MESBIC).

"Actually, we went in three different directions to raise money," Holley recalls. "The first was the MESBIC, which would provide a loan, basically; it would be a very soft debt, with a ten-year pay-back

period and low interest." In addition, Mary Anne notes, they searched for individual investors, one of whom provided funds for a mail survey that showed an excellent market for the magazine. The third source of assistance for the magazine's start-up is a line of credit with suppliers which would allow the Holleys ninety days to pay the printing and paper firms. The couple has located a bank that will guarantee this credit arrangement with suppliers, thereby lessening the cash needed for the initial publication.

One reason for this involved funding process is the million-dollar budget necessary to start a magazine, and another complicating factor is the MESBIC investment stipulations, Mary Anne explains. "If you have MESBIC financing, you have to have

at least 51 percent minority ownership—a combination of us and the investors together." The Holleys have put together an investment package that meets these requirements, although they must wait until interest rates are lower for the loans and investments to come through.

During the early stages of this search for funding, Mary Anne worked at home full-time on the applications and details of financing. For the past year, however, she has been getting invaluable experience as an assistant to publishing consultant James Kobak, working from his New York workstead office. "I couldn't have created a better job," Mary Anne states. "I wanted to expand the experience I've had in our own magazine project, and I'm able to do that here."

grams. The most common is the "7A" loan guarantee, whereby the SBA guarantees 90 percent of a bank loan for a business or, in some instances, loans the money directly if the applicant has been turned down by two banks. These "7A" loans require that the borrower put up 50 percent of the total capital needed, however.

Two other SBA programs, Operation Business Mainstream and Economic Opportunity Loans, guarantee up to 80 percent of a bank loan to minority businesspersons, the economically disadvantaged, and handicapped persons. Another SBA plan guarantees a seasonal line of credit for qualifying businesses; this program provides short-term loans for labor and materials in a firm's busiest work times and requires pay-back within a year.

The federal government also sanctions Small Business Investment Companies (SBICs) and Minority Enterprise Small Business Investment Companies (MESBICs), which combine private funds with SBA money to invest in and make loans to qualifying firms. Like SBA or conventional bank loans, funding from these investment companies requires comprehensive application data and months of waiting.

Comparatively limited funds are available to artists and arts-related businesses through grants from the National Endowment for the Arts. (The similar National Endowment for the Humanities also awards grants, but its qualifying categories are less amenable to workstead businesses.) Although these grants provide ideal seed money for some worksteaders, they generally are limited to subsistence-level funding for only one year.

Many state and community programs also support small business ventures that benefit their communities. Often these programs utilize federal funds and local resources, as does Baltimore's "shop-steading" project. In this unusual experiment, businesspeople can buy abandoned commercial buildings in a rundown section of the city for $100. With help from city officials, the seventeen buyers who joined this project have found low-interest loans to renovate the buildings and install their businesses, which include a pottery-making enterprise, a stained-glass production workshop, and a shoe store—all of them worksteads.

Applying for Loans

For government or institutional loans or grants, there is no way to avoid the lengthy, complex application process. Yet worksteaders can improve their chances of success and perhaps discover new sources of funding by asking questions of other small-business owners and all the bankers and bureaucrats they can reach. This personal research will identify specific programs and institutions that are appropriate for a particular workstead's needs.

Various reference books on small business and finance include general surveys of loan sources, but their greatest value to a work-steader's capital search is the sample loan-application packages and business plans they contain. Comparison of several such documents

Janice Mae Schopfer:
A Stable for the Book Arts

Janice Mae Schopfer shares an old San Francisco stable with a roommate and the headquarters of the Pacific Center for the Book Arts, a group that she helped to found. Her own work includes conception, design, illustration, and binding of small-edition works of contemporary literature, along with some mounting of prints for museums and galleries.

For the past nine months, Janice's work and study have been financed by a grant from the National Endowment for the Arts. "This is one of the first times the book arts have been recognized as a craft," she notes. "In the past, these grants have gone to fine printers from a fund for literature, not for crafts." Her award, called an apprentice-ship fellowship, allows Schopfer to study with a master teacher of bookbinding, the phase of bookmaking that Janice recently has added to her work as an illustrator and apprentice printer. "In my case, the grant is paying my tuition to study with this teacher, plus it pays for my materials and my rent."

Schopfer helped to organize the book arts center, which occupies the downstairs of her building, to encourage cooperative ventures in producing fine books and to provide space for some of the heavy equipment used in making books by hand. Janice believes that contributing time and energy to this effort will benefit everyone working in the book arts. "In order to get things moving, you have to sacrifice—to do work besides your own pieces, so that you help to improve the scene."

Schopfer's successful application for an NEA grant is evidence of "improvement of the scene" for book artists, and she enjoys living where much of this activity takes place. The Center's equipment supplements her own, which is housed in the hayloft of the stable, along with her antique brass bed, a cat, and a comfortable array of furniture and plants. "I like my work to be close at hand; I can work on it any time. If I'm working at something, I can stay with it until I finish and then walk ten paces to my left and fall down and sleep."

will enable an applicant to identify all the categories of information that are pertinent to her business and financial situation.

In many urban areas, community organizations provide free consultation and financing assistance to small-business owners. The SBA or a local Office of Minority Business can provide lists of these organizations, which are not always limited to assisting minority businesspeople. A number of business-development organizations, such as the Baltimore group that initiated the shopsteading program, focus on particular urban areas and offer help to everyone interested.

Some private financial consultants also specialize in assisting small businesses. These experts either charge a flat fee or take a "finder's fee" that amounts to a small percentage (5 percent or less) of the capital they raise for a venture. A worksteader should be cautious in selecting a private consultant, however; bankers, accountants, or other finance professionals probably can suggest the names of appropriate consultants for a workstead enterprise.

After information and advice have been garnered, the worksteader is ready to take the essential steps toward raising capital on her own. Using loan applications and reference books as guides, she must develop all the elements of a business plan, budget, and personal financial statement that will show any lender that the workstead enterprise and its proprietor are competent, worthy recipients of the lender's trust and—more important—its money.

Planning and Recordkeeping

The basic loan-application process is an excellent means of evaluating a workstead's health and prospects, even if it already is a going concern or doesn't require capital from outside sources. If this evaluation is made annually (whether or not the worksteader intends to borrow money), the owner can readily see the growth as well as the strong and weak spots in the business. At the very least, a home-based businessperson should keep careful track of his financial status; the chart on page 131 suggests one possible schedule for this evaluation.

Setting Goals

Just as important as a plan that details how the business will operate is a set of goals for its development. Although this concern may seem alien to worksteaders who prefer day-to-day flexibility in their livelihoods, every businessperson should know the minimum sales or billings necessary for the venture's survival, and projecting this work over a period of months or years helps build control into the operation. Similarly, a worksteader should project personal and family living expenses for several years into the future, so that she can set income goals for her independent business. (See page 133 for guides to estimating both personal and business expenses.)

Small-Business Financial Status Checklist

(What an Owner-Manager Should Know)

DAILY
1. Cash on hand.
2. Bank balance (keep business and personal funds separate).
3. Daily Summary of sales and cash receipts.
4. That all errors in recording collections on accounts are corrected.
5. That a record of all monies paid out, by cash or check, is maintained.

WEEKLY
1. Accounts Receivable (take action on slow payers).
2. Accounts Payable (take advantage of discounts).
3. Payroll (records should include name and address of employee, Social Security number, number of exemptions, date ending the pay period, hours worked, rate of pay, total wages, deductions, net pay, check number).
4. Taxes and reports to State and Federal Government (sales, withholding, Social Security, etc.).

MONTHLY
1. That all Journal entries are classified according to like elements (these should be generally accepted and standardized for both income and expense) and posted to General Ledger.

2. That a Profit and Loss Statement for the month is available within a reasonable time, usually 10 to 15 days following the close of the month. This shows the income of the business for the month, the expense incurred in obtaining the income, and the profit or loss resulting. From this, take action to eliminate loss (adjust mark-up? reduce overhead expense? pilferage? incorrect tax reporting? incorrect buying procedures? failure to take advantage of cash discounts?).
3. That a Balance Sheet accompanies the Profit and Loss Statement. This shows assets (what the business has), liabilities (what the business owes), and the investment of the owner.
4. The Bank Statement is reconciled. (That is, the owner's books are in agreement with the bank's record of the cash balance.)
5. The Petty Cash Account is in balance. (The actual cash in the Petty Cash Box plus the total of the paid-out slips that have not been charged to expense total the amount set aside as petty cash.)
6. That all Federal Tax Deposits, Withheld Income, and FICA Taxes (Form 501) and State Taxes are made.
7. That Accounts Receivable are aged, i.e., 30, 60, 90 days, etc., past due. (Work all bad and slow accounts.)
8. That Inventory Control is worked to remove dead stock and order new stock. (What moves slowly? Reduce. What moves fast? Increase.)

From "Keeping Records in Small Business," by John Cotton, Small Marketers Aid No. 155, Small Business Administration, Washington, D.C., 1974.

Even if these predictions of income and needs are ultimately inaccurate, worksteaders will benefit from the process of examining their enterprises. As Bill Welch of Glad Hand Designs recalls, "We made projections for each month's sales and production before we started. We didn't meet them, but now we know how to plan realistically."

Taking Stock

It's easy for an independent professional to get so absorbed in his work that he forgets to look where the business is going. But checking the present status of accounts, supplies, and clients' needs is an important part of business planning, especially for worksteaders who may be somewhat out of the mainstream of activity in their occupations. Taking the pulse of the business also includes consulting employees and planning to remedy any problems they have noted, as well as talking with family members about making the workstead business compatible with their needs and expectations. These business-reviewing activities can be done informally, although they are most likely to be accomplished if the worksteader sets aside regular times for them.

Keeping Records

Worksteaders should keep complete records and copies of all transactions and correspondence for at least several years. The filing system need not be elaborate or costly, but the stored documents should back up the more abbreviated records of a ledger or order book. Then if a client questions a charge or changes an order, the original materials can be consulted and the problem ironed out. For example, Mike Young, whose computer mailing list service stores a priceless asset for customers, has instituted three separate backup systems for his clients' records, so that if the computer goes haywire he can reconstruct the information.

Financial records are the heart of many businesses, and if a workstead is subject to a tax audit or investors' examination, they could spell life or death for the venture. Though their importance is clear, a beginning businessperson (or long-established one, for that matter) may pale before the ledger pages. With the help of an experienced bookkeeper or any of several good reference books, however, a simple and accurate accounting system can be established. The precise method or arrangement of the recordkeeping is less important than its features—it should be easy to use and understand, consistent, and up-to-date.

Accountant Nancy Feiner suggests that worksteaders use a simple ledger that includes all business and personal expenses. She maintains that one checking account is adequate for both purposes, so long as the worksteader keeps careful records of receipts and expenditures. Other home-based businesspeople prefer to have separate

Money Needs for First Three Months of a Business

LIVING EXPENSES	From last paycheck to opening day	$ _____
	Moving expense	_____
	For three months after opening day (from cost-of-living budget)	_____
DEPOSITS, PREPAYMENTS, LICENSES	Last month's business rent (first three months in operating expenses below)	_____
	Telephone and utility deposits	_____
	Sales tax deposit	_____
	Business licenses	_____
	Insurance premiums	_____
LEASEHOLD IMPROVEMENTS	Remodeling and redecorating	_____
	Fixtures, equipment, displays	_____
	Installation labor	_____
	Signs—outside and inside	_____
INVENTORY	Service, delivery equipment, and supplies	_____
	Merchandise (approximately 65% of this amount to be invested in opening stock)	_____
TOTAL OPERATING EXPENSES FOR THREE MONTHS (from projected profit and loss statement)		_____
Reserve to Carry Customers' Accounts		_____
Cash for Petty Cash, Change, etc.		_____
	TOTAL	$ _____

Money for living and business expenses for at least three months should be set aside in a bank savings account and should not be used for any other purpose. This is a "cushion" to help get through the starting period with a minimum of worry. If expense money for a longer period can be provided, it will add to peace of mind and help the entrepreneur concentrate on building the business.

*Personal Cost-of-Living Budget**

Regular Monthly Payments

Rent or mortgage (including taxes)	$ _____
Car (including insurance)	_____
Appliances/TV	_____
Home improvement loan	_____
Personal loan	_____
Health plan	_____
Life insurance premiums	_____
Other insurance premiums	_____
Miscellaneous	_____
TOTAL	$ _____

Household Operating Expenses

Telephone	$ _____
Gas and electricity	_____
Water	_____
Other household expenses, repairs, maintenance	_____
TOTAL	$ _____

Food Expenses

Food—at home	$ _____
Food—away from home	_____
TOTAL	$ _____

Personal Expenses

Clothing, cleaning, laundry	$ _____

Drugs	_____
Doctors and dentists	_____
Education	_____
Dues	_____
Gifts and contributions	_____
Travel	_____
Newspapers, magazines, books	_____
Auto upkeep, gas, and parking	_____
Spending money, allowances	_____
TOTAL	$ _____

Tax Expenses

Federal and state income taxes	$ _____
Personal property taxes	_____
Other taxes	_____
TOTAL	$ _____

BUDGET SUMMARY

Regular Monthly Payments	$ _____
Household Operating Expenses	_____
Food Expenses	_____
Personal Expenses	_____
Tax Expenses	_____
MONTHLY TOTAL	$ _____

*This budget is based on an average month. It does not cover the purchase of any new items except emergency replacements.

Reprinted with permission from Bank of America NT&SA, "Steps to Starting a Business," *Small Business Reporter,* Vol. 14, No. 7, copyright © 1980.

bank accounts for home and work, often at different banks so that statements arrive at different times of the month and checks can be easily identified for each purpose. Every worksteader will find an adaptation of standard recordkeeping that suits his needs; the crucial element is to use that system consistently.

A useful booklet published by the SBA, "Keeping Records in Small Business," offers a summary of basic recordkeeping needs. It lists four essential types of records: sales, cash receipts, expenditures, and accounts receivable. These categories can be subdivided according to the specific requirements of a particular firm. Other necessary records include a list of all business equipment, a schedule of depreciation for major items, a record of insurance coverage, and payroll records. Several bookkeeping systems that cover all of these categories are available in office supply stores, or a worksteader can easily make her own ledger sheets and other record forms.

Taxes

As with many other aspects of doing business, the applicable taxes and the strategies for handling them vary with both occupation and location. For the inevitable one, though—federal income tax—worksteaders should be sure to take advantage of tax savings that are specific to working at home. Many of these advantages apply to salaried persons who use a home office exclusively, as well as to self-employed businesspeople. (See page 139 for a list of typical expenses that apply to all businesses.) The discussion of tax deductions relating to home offices found on pages 137–142 was prepared by accountant Nancy Feiner, whose five hundred clients include a majority of workstead businesspeople.

Income Tax

The federal government and most states levy income taxes that usually must be estimated and paid quarterly by self-employed persons, which includes most worksteaders. (These taxpayers still must file a yearly return.) The state forms and regulations generally follow the federal ones, so at least the work of wading through complex instructions and piles of records need only be done once.

Social Security

All self-employed persons must pay a federal self-employment tax, even if they have another type of retirement fund as well. The current rate is 8.1 percent of the first $22,900 of income, which is higher than the rate for salaried employees; this tax is slated to increase for everybody for the 1981 tax year. Self-employed persons do qualify for Keogh Plans, a form of tax-deferred retirement account. Contributions to a Keogh account are not subject to income tax, but they must be included as income when self-employment tax is calculated.

Robin Davey and Bill Welch:

Wine Barrels to Potholders

Financial considerations figure prominently in the operation of Glad Hand Designs, Robin Davey asserts: "If we continue to do things right—and the economy doesn't go all to hell—Glad Hand should be worth a good deal in two years." Davey and her husband, Bill Welch, are founders and co-owners (with an investor friend) of this kitchen accessories business, and they have definite plans for the growth and eventual sale of their company. Just three years old, Glad Hand has six employees in addition to Robin and Bill, and turns out an annual total of 20,000 potholders, aprons, and tea cozies. All have silk-screen designs by Robin, Bill, or one of Robin's artist

parents, and most are produced in a small brick warehouse that once served as a cooperage for local winemakers. The sewing on some items is done by the three employees who work in their own homes.

At present, Robin and Bill are coping with the rapid growth of their venture. They have nationwide sales representation, and the volume of orders recently forced them out of a spacious six-room apartment and into their cooperage workstead. Although Robin describes Glad Hand's growth cycle as "sort of like a dog chasing its tail," both agree on their objectives for the company. As Bill notes, "We put ourselves in the position of having

to make money; we hired people, then worked extra hard to get the cash flow to support them."

Employee Taxes

If a worksteader hires employees, federal and state income tax and Social Security must be withheld from their salaries and a number of related forms filed at various times of the year. The IRS's "Employer's Tax Guide" provides instructions for these procedures. In addition, employers may have to contribute and withhold sums for federal and state unemployment insurance and disability coverage.

Property Taxes

Worksteaders who own their homes or other buildings are no doubt familiar with this form of tax. Some communities may give tax breaks to worksteads, such as an exemption from property-tax reassessment after an old building is renovated.

Business-related Taxes

These taxes vary with the type of business and its location.

Sales tax. This state tax (and, in some places, a city tax as well) is levied on most retail items. Worksteaders who sell retail goods must obtain a permit (for a small fee) and establish an account (as much as $300) against anticipated sales tax that will be collected. Sales tax usually is paid quarterly, and many advisors recommend that businesspeople set up a savings account for sales taxes so that the money will be available when due and will be earning interest in the meantime.

Gross-receipts or inventory tax. Some states and localities charge a small percentage tax on yearly receipts or the value of business inventory on a given date each year.

Excise or individual-item taxes. Certain commodities are taxed for being what they are, such as alcoholic beverages and tires (which carry a federal excise tax), or timber, soft drinks, and playing cards (which are taxable in certain states).

Reference Guides from the IRS

The following booklets provide details on specific obligations and are free for the asking at any IRS office:

Your Federal Income Tax (Publication 17)
Tax Guide for Small Business (Publication 334)
Business Use of Your Home (Publication 587)
Employer's Tax Guide (Circular E)
Self-Employment Tax (Publication 533)
Tax Information on Retirement Plans for the Self-Employed
   (Publication 560)
Tax Information on Depreciation (Publication 534)
Information on Excise Taxes (Publication 510)

The Home Office and the IRS
by Nancy Feiner

Like it or not, taxes are an integral part of life, and while someone might choose to work at home to make life more pleasant, if that person has net income, then income taxes have to be paid. Our government has a vested interest in helping citizens succeed in business, and so a home businessperson's savings today will be paid back in increased taxes based on bigger profits tomorrow, and in the meantime cash outlay is lessened.

There are significant tax savings available to people who work at home in the form of tax deductions, tax credits, and depreciation allowances. The information in this section comes from my work as a licensed tax preparer in California. While most of these suggestions may be used by any taxpayer, every working person's tax situation is different, and each person must take full responsibility for his or her tax return and business records. In addition, tax rules change to some extent every year, so taxpayers should consult up-to-date reference sources or reputable tax preparers for detailed advice.

Furnishings and equipment. Working in your home allows you to begin operations with very little capital and, as the business grows and profit is made, to make the necessary equipment purchases gradually. Additionally, there are great income tax advantages available just when you need them. Working at home gives you the opportunity to put items you already own into business use and receive a deduction for depreciation. (Depreciation is an allowance for loss in value as an item is used; for example, a $250 camera might be "depreciated" for tax purposes at $50 per year for five years.) While you no longer can use these items for personal reasons (unless you apportion a percentage for personal use that is not deductible), you are saved the cash outlay for these items and also save money on your income taxes. For instance, you may decide that there are some chairs, tables, and wall furnishings in other rooms in your home which you can utilize in your office. You may take the lesser amount of their purchase cost or used value and list those items on your depreciation schedule (see below). You must depreciate these items because (1) they probably still have a useful life of more than one year, and (2) they were purchased prior to the current taxable year.

Similarly, self-employed persons working at home also can deduct the tools, books, journals, and other equipment that they purchase for business use. You should take a physical inventory of these items, determine their value (if purchased prior to the current taxable year, you must value them at the lesser of used value or purchase cost), and set up your depreciation schedule. Look around your home and see what you are using and forgetting to deduct. Suppose you give piano lessons or are a musician. Remember that all or part (depending on your usage) of that stereo you listen to is just as deductible as the musical instrument you play. And what about your record library? If it is used professionally, it should be valued, too. If you run a child day-care facility in your home, the IRS has special provisions if you have a state license: you may take a proportionate share of your kitchen and other home facilities—dishes, pots and pans, washer and dryer, and other equipment—even though they are not used exclusively for business.

Computing depreciation. Determine the shortest life that is reasonable for an item within the IRS guidelines. For instance, a screwdriver that cost $5 might last ten years, but it would be unrealistic for you to depreciate it for ten years. First of all, consider the amount. If an item's value is less than $100, you might want to deduct its full amount in the current taxable year. If the value exceeds $100 (cost of new item or value of used item), then base your decision of how long to depreciate it on the useful life of the item to you—not how long you think it might last.

Investment credit. There is a tax credit available to you for any item purchased (new or used) in the current year for business purposes. This tax credit is allowed *in addition* to any deduction for depreciation on an item with a useful life of three years or more; the amount of the credit is dependent on the number of years you depreciate the item, but can be no more than 10 percent of the cost of the item. Like other tax credits, the amount of the investment credit is subtracted from taxes owed, rather than from gross income.

Office or work space. One deduction that involves no immediate cash outlay is the office or work space in your home. You may deduct this expense when you have a space *exclusively* set aside for the use of your business or profession *and* when *no other* space has been provided for you by anyone else. If you have been provided office or work space elsewhere, then you cannot deduct office space at home, because this is merely for your convenience. In a situation where you are paid as a self-employed person and have an office at home, but also have work space available at the place of the person who is paying you, you are likewise not allowed the home office deduction. However, suppose you are a professional subletting someone else's office on an hourly or time basis and are limited in the availability of that office to you. Because of that, you work at home also. In this situation, you may take a home office deduction on the percentage of time you use your home work space.

How do you figure the home work-space deduction? To determine the amount of space deductible, measure (walking it off is reasonable) the total square footage in your living space (include the garage if you are using that for business purposes) and the total square footage in your work space. You then arrive at a percentage by dividing the work space by the total space.

The work space must be used exclusively for your business or profession, but if your work requires storage space as well, you may include this in your percentage computation. The storage space does not have to be used exclusively for business purposes. For example, if you have a space in your garage or closet where you

Typical Business Expenses

(Deductible from Gross Income)

Account books
Accounting fees
Advertising
Auditing fees
Automobile expenses
Bad debts
Bank service charges
Bonding fees
Bookkeeping services
Books
Burglar alarm service
Business associations
Business cards
Business gifts
Business interruption
   insurance
Business license
Casualty losses
   (robbery, fire, theft,
   burglary, shop-
   lifting, vandalism)
Charitable
   contributions
Cleaning
Clothing, special
   (not regular street
   clothes)
Coffee service
Collection expense
Commissions

Consultants' fees
Contractors' fees
Conventions
Cost-of-goods-sold
Credit bureau fees
Credit card fees for
   merchants
Depreciation
Dues, business
   associations
Dues, professional
   societies
Dues, union
Education expenses
Electricity
Employer's taxes
Employment agency
   fees
Entertainment
Equipment
Extended coverage
   insurance
Fees for services
Fees to professional
   organizations
Fire insurance
Freight
Garbage
Gross-receipts tax
Interest on business
   debt

Inventory
Inventory tax
Janitorial service
Ledgers
Legal expenses
Liability insurance
License fees
Loss on sale of
   business assets
Machinery
Magazines
Merchants'
   associations
Minor repairs
Moving expenses
Night-watch service
Office furnishings
Office in home
Office supplies
Passport fees for
   business trip
Patents
Payroll and withheld
   payroll taxes
Periodicals
Permit fees
Postage
Professional fees
Professional journals

Property taxes
Publications
Reference books
Rent
Repairs (subject
   to certain
   requirements)
Research and
   experimentation
Safe deposit box
Salaries
Sales tax
Service charges
State income tax
   (not deductible on
   state return)
Stationery
Supplies
Telephone and
   telegraph
Theft insurance
This book
Tools
Trademarks
Travel away from
   home
Uniforms
Unincorporated
   business tax
Utilities

Adapted from *Small-Time Operator* (Revised Edition, 1980), by Bernard Kamoroff, Bell Springs Publishing, Laytonville, CA.

have other things stored, simply figure the amount of space your business equipment takes up.

If you're a renter, you compute the deduction for home work space by simply multiplying your rent by the business percentage. Many people ask how they can figure the space if they have a studio apartment or use part of their living room for work. Furniture arrangement is the key here. The exclusive space must be created to satisfy the IRS, and that work space cannot be used for personal purposes. So create a separate work area with placement of furniture, screens, or other devices, which not only will satisfy the requirements of the IRS but also will lend a more businesslike atmosphere. While clients have come to appreciate more and more the informality of home offices, it is important not to be too informal and to set some business standards from the beginning.

If you own your home, you determine your work-space allowance by depreciating the house: divide the building cost (land cost is not deductible) by years of useful life and multiply that figure by the business percentage. There is, however, a way to increase your tax savings for home work space if your office is located in a new addition to your home. Depreciating the addition only, rather than the entire house, often results in a larger deduction. Suppose, for example, that the basic cost of your house and lot was $30,000; the addition cost $20,000; and the cost of the lot alone was $15,000. Suppose, too, that the percentage of office space to the total space of your house is 25 percent. Using the "whole-house" method suggested previously, first you must subtract the cost of the lot ($15,000) from the total cost of the house ($50,000) because land cost is not tax-deductible. Then you depreciate 25 percent of the $35,000 for, say, twenty-five years and take an annual deduction of $350. However, if you just depreciate the addition (in which case you would not have to deduct land value), you set up your depreciation schedule for $20,000 over twenty-five years, which would give you a yearly deduction of $800.

Home maintenance and improvements. You can charge some of the upkeep of your home to business use; thus, tax savings enable you to keep your home looking good for less than it ordinarily would cost you. Yard maintenance is essential if you see clients in your home, and the cost of landscaping may even be deductible for business purposes. If your business requires special electrical or plumbing hookups, or the addition of other kinds of fixtures, these costs are deductible also. (If you make expensive improvements, which increase the value of your home, you may depreciate them.)

Utilities, insurance, and other deductibles. Your percentage-of-work-space-to-living-space formula also applies to utilities, insurance, interest, and taxes. Remember to apportion the interest and taxes between work use, on Schedule C (business income), and personal use, on Schedule A (itemized deductions) on your tax return. By taking the proportionate percentage on Schedule C, you can reduce your self-employment tax. You are entitled to take these deductions in all cases on Schedule A, but it can be of added use as part of your office in the home.

The cost of home utilities and insurance is not deductible except for business use, so the list of tax deductions for home business-people keeps increasing. The deduction for business use may be as much as 50 percent of your utility bill, but you will have to compute this and be able to substantiate it to the IRS. This also may be true of your insurance. The cost of any special business coverage should be deductible, as well as the business percentage of basic household insurance premiums.

Telephone. The telephone is *not* a utility. You cannot apply your office-in-the-home percentage to the telephone; the IRS allows only business use of a phone as a deduction. If you have one phone which you use for business and personal purposes, then the base rate is not deductible. The only deductible items are business-related toll calls, long-distance calls, and other special-rate services, such as a WATS line or tie line. To keep track of business calls, you might want to note the breakdown between business and personal charges on the front of every bill. If you want to have a second phone line installed in your home, a line that you allocate exclusively for business use, then you can deduct 100 percent of this phone bill and continue to deduct the business-use portion of your personal phone.

Automobile expenses. When a person is working at a salaried job, none of the expense of getting to and from work is tax-deductible. When a self-employed person is working out of an office, then the use of an automobile is deductible only after that person checks into the office. But for the person working at home, deductible mileage begins whenever that person drives away from the house to do work-related business.

Most people tend to underestimate their business mileage rather than overestimate it. Many people who work at home simply forget to count all the times they run out to do errands or pick up supplies, because they are leaving from and returning home and haven't developed a business attitude related to income tax savings. Perhaps you have to see one of your clients to pick up an invoice, and on the way, in a relative straight line, you decide to save time by stopping at the grocery store. Your principal intent in leaving your home was to conduct a business transaction; therefore the total mileage is deductible. And what about the times you go to the bank to make a business deposit or to the post office or stationery store? Remember to count the miles—they really add up.

There are two methods of keeping records to maximize auto deductions, and every home businessperson should keep both. One method is to keep a record of auto maintenance expenses (regular method), and the other is to deduct the current cents-per-mile allowable by the government (optional method). In each case, only work-related auto usage is allowed, so the appropriate percentage of total car usage must be determined. You can do this by keeping a reasonable record of business driving; your appointment book is a great place to keep track, or you can purchase a printed mileage book at any stationery store.

If you are keeping a record of maintenance expenses, you must have receipts or canceled checks for maintenance and insurance payments, but gas receipts are not necessary. If you have been keeping a mileage record, you can easily figure the amount of money you spend on gas per year by using the following formula:

$$\frac{\text{Total miles driven in year}}{\text{Miles per gallon your car gets}} \times \begin{array}{c} \text{Average price} \\ \text{of gas per gallon} \end{array}$$

You then add your total gas cost to your maintenance, repair, and insurance costs plus the depreciation value of your car and multiply this figure times your business-use percentage. Compare this figure with the current cents-per-mile times business miles and take the larger deduction. If you are keeping good business records this is not complicated, since all you have to keep track of are maintenance expenses, your beginning and ending speedometer readings for total mileage, and business miles (your best estimate).

There is one contingency that must be mentioned. Let's say you discover that the regular method gives you a much greater write-off since you haven't driven many business miles and you either own a new auto and have a large depreciation or have put many repairs into your older car. You set up a schedule depreciating your auto at, say, three years. The second year, however, you discover that the cents-per-mile method is more advantageous. The IRS has built depreciation into the cents-per-mile figure, so it counts this as your second year of depreciation. The third year you decide to switch back to the regular method. The IRS now considers your three-year depreciation used up, so that in the future, if you still have the same car, you will be allowed your maintenance expenses but no depreciation using the regular method, or only ten cents a mile for the optional method. If you always choose to use the optional method (cents-per-mile), you may use it endlessly—there are no limitations or depreciation problems.

Another situation: suppose that you have been using the regular method and have used up the depreciation on a car. If you have two automobiles, you could retire the depreciated car from business use and begin using the second auto for business purposes; this means that you can set up a new depreciation schedule for the second car.

Think taxes. In summary, you should develop a business consciousness when working from home and constantly think whether each purchase or transaction has an income tax advantage. It is important not to forget this in the informal working-living situation of your home.

Nancy Feiner:

A Relaxed Approach to Taxes

Accountant Nancy Feiner founded her workstead business seven years ago, doing bookkeeping and tax returns for about seventy-five clients. Since then her business has grown to more than five hundred clients, and she has put a two-room office addition on her home. During the busiest four months of the year, she employs three assistants.

A divorced mother of three children, Nancy has developed a home business that supports her family, who can be an occasional source of disruption while she's working. "You worry about your house being clean, or the kids watching TV or playing loud music. The other night my son came bounding in to tell me something, and I had to remind him that I was with a client. But it's not the family's fault most of the time," Feiner adds. "They have a right to live in their house."

On balance, the accountant prefers the compromises and interruptions to an office elsewhere, and she is confident that her clients do, too. "People like coming to me in my home; they tell me that constantly," Nancy states. "They feel that they can be honest—they can tell me whatever is going on and I will tell them how to approach it, what's legal and what's not. They're not intimidated by my presence, and they're not put off as they might be in a colder sort of business office."

Feiner also notes that by chatting with people casually, she often learns background information that will save them money on taxes: "By talking to people here, I find out so many things that I'd never have known if they'd just come in and picked up their forms. I think that happens because I'm working in my home and because the atmosphere is relaxed."

Priorities

Even though priorities seem to change every time we write them down, most of us pursue this ordering of the important goals and processes in our lives. For worksteaders, there are some specific priorities that govern the functioning of a home-based business: these are routine, the mechanics of getting at the job; time management, the struggle with the clock; and morale, the intangible and delicate province of mind and mood.

Routine

Every worksteader is subject to certain "hazards" inherent in working at home. Sometimes these hazards are also our pleasures—family, friends, phone calls from old roommates or distant cousins—while at other times the hazards are more mundane—the garden needs weeding, the piano is dusty, the refrigerator beckons. During the course of a work day, these common distractions can disrupt the functioning of a workstead business. With some planning and educating, however, the worksteader can minimize both the temptations and the interruptions.

Educating Friends and Family

Perhaps the most frequent complaint of home-based professionals is that the people around them don't take their work seriously. Variations on, "Well, you're at home, so you couldn't really be working," are reported by most worksteaders, particularly those just beginning home careers. "People know you're at home," notes publicist Candice Jacobson, "so they just drop in. Usually I tell them I'm working and ask them to come back after five o'clock or to call so we can get together for lunch."

One or two polite turndowns should discourage such unexpected work-time visitors, but a worksteader can avoid straining personal relationships by educating her family and friends about her work schedule and demands. Of course, this means that she herself must take the work seriously, and this attitude will help her persuade others to make social calls or visits after business hours. Her educational efforts also will help her organize her work days, because she's likely to be asked about the details of starting and quitting times and the possible exceptions to her rules.

Some friends or family members may be offended by this necessary separation of work and leisure hours, but these people probably will accept the distinction in time. The wife of a medical researcher makes a mental adjustment to her husband's home work schedule: "When he's in his lab, I just pretend he has gone to an office somewhere else—it frees both of us from the daily business of getting on each other's nerves."

Don McClelland:

Writing on a Solar Schedule

Poet and novelist Don McClelland has found that the sun determines his work schedule and his choice of work area: "I'm a heliotropic person—I know where the sun is all the time, and I'm very aware of light. The kind of light that falls on the page that I'm writing is very important to me." Don works in an east-facing room to take the fullest advantage of the morning sun, and he's a purist about it. "I tried having curtains on the windows in the room," he states, "but I didn't like the light that came through them."

McClelland rises with (or before) the sun and is usually at his desk by 5:15 A.M. He gets in two hours of work before his wife, Pat, and daughter, Ula, get up, and then joins his family for breakfast. "I just hang out with them until they leave at about eight thirty. That's a dynamic," the writer stresses, "not an administrative task. I want to be with them because I know I'm not going to see them for the rest of the day."

With the house to himself, Don resumes writing or researching for his fiction. And this time is the source of distraction for him: "When the neighborhood begins to come alive—about nine thirty, usually—the temptation is the worst. The very fact of enjoying yourself working in a domestic environment means that you're enjoying your surroundings, too—your friends on the block and the kids in the park at the end of the street. So when I decide to give in to temptation, I try to do it in a programmed way, to let myself do one thing, like walking to the park to see who's on the swings, and then coming back to work."

McClelland has achieved a good routine of spending time with his wife and daughter and taking breaks when working alone, but he did have to establish some ground rules when he started his workstead career. "We rotate a lot of the domestic jobs," Don notes, "but because you're home, you can get stuck with a number of jobs. Your spouse calls up and says,'Hey, as long as you're home, why don't you put a little laundry in and I'll take care of it when I get home.' So we decided that there would have to be almost ironclad rules, so that if Pat calls up and asks me to do the laundry, then I get to call her up and ask her to write my book. Of course that hasn't happened, but we've all come to understand that writing is a full-time job. That's the way it has to be, because otherwise a studio at home is no sanctuary—it's a trap."

Another nerve-jangler, the telephone, is a source of interruption that can be handled in several ways. In addition to telling friends to limit personal calls to nonbusiness hours, the worksteader can save a lot of time by keeping work-related calls brief. Some worksteaders use an answering machine to screen calls during business hours, and others have an employee take messages when they're doing concentrated work. One woman answers her phone with her business name during the day and "Hello" after six at night, a clear signal that she does not want to discuss business in the evening.

Self-discipline

Just as worksteaders must educate their families and friends to observe business hours, they also must discipline themselves to work during those times. And for many of us, getting down to work is much harder than silencing the doorbell and the telephone. This is the individual struggle with the papers or the tools or the ideas that is the challenge of all work, and of everyone who pursues it.

The individual struggle is personal, certainly; but the exercise of discipline can be enhanced by a systematic approach to working. The collective experience of a hundred worksteaders suggests that goals, rituals, rewards, and contingencies are the key elements for a good system of discipline. Together these elements work to involve the worksteader so thoroughly in his business activities that goofing off doesn't even occur to him.

Goals are necessarily specific to each occupation and person, but for purposes of daily discipline they should be realistic and short-range—projected, say, over a few days or a week. Ideally, a worksteader sets his next day's goals at the end of the present day, or at a fruitful point in the work, because then he won't waste half the next day trying to decide what to do. This might be called the Hemingway Axiom, after the novelist's advice to young writers to stop each day's work when it's still going well so that they'll know right where to start in the morning.

Rituals are also personal, but one that is common to many worksteaders is going out for coffee. Architect John Edwards prefers "going through the weather" to make a clean distinction between leisure (or "living") and work, so he jogs or takes his dog for a brisk walk each morning. Others may not leave their dwellings, but simply perform a small, symbolic act to initiate the work day, such as sharpening pencils, pulling up a shade in the office, or taking a cup of tea to the desk.

Publisher Malcolm Margolin has a ritual to begin his week: each Monday he walks to the post office and collects the mail for his company, Heyday Books. He enjoys this bit of physical activity, and the contents of his post office box essentially order his week's work.

Rewards should be built into each work day. Usually they involve time off from work, such as lunch with friends, a short game of catch with the children after school, or an occasional matinee movie or afternoon at the beach. A worksteader has to invent these small daily rewards because she's seldom likely to be in an economic position to give herself a raise or a promotion. She can, however, award herself—and her co-workers, if appropriate—with a day off. Whether the reward is a ten-minute personal phone call or a twelve-dollar lunch, the effect is usually the same—to give a lift to the day and replenish a worksteader's energies.

Yet there are times when our energy or our effectiveness seems to disappear, and the usual tricks won't bring it back. This is when a contingency plan should be put into operation—even if the "plan" is entirely impromptu. A contingency plan does not mean the worksteader bails out altogether, however. For example, anyone who does typing as part of his job has days when his fingers are completely out of synch with the machine; at such times, it's more productive and far less frustrating to abandon the typewriter for a few hours and concentrate on some other tasks.

Generally, a change of pace will "cure" a motivation problem, but in some instances a more radical solution may be necessary. Photographer Paul Caponigro took such a radical step when he informed all the galleries and museums that had purchased his photographs that he would not be printing any of his "old" work for two years. Caponigro felt compelled to take this action because "all the demands on my time were for old work, and I had no time for the new. I felt that I wasn't growing." This break with the "old" accomplished its purpose for Paul, who freed himself to concentrate on a new series of prints in his workstead darkroom and to assemble a new body of work that is now being shown along with his previously acclaimed photographs.

Domestic Tranquility

Another aspect of establishing a workstead routine is to designate clear boundaries for the living and work areas, as well as the housekeeping chores for each. A number of worksteaders have pointed out that bits of their work paraphernalia spill over into the living quarters, which can create a sense of the work's being out of control. When Robin Davey and Bill Welch shared their six-room apartment with their kitchen accessories business, for example, both were zealous about observing their rule that no fabric or thread or paperwork from the business could be brought into the living areas.

Similarly, most worksteaders point out that doing business in their homes requires that they keep things neat and clean. Lofts or one-room studios obviously merge office and living areas, so that any messy corner will be apparent to visitors and workers. Even in residences where the office is located in a separate room, living areas may be used for meetings or private consultations. Ernest and Anita Scott of San Francisco Book Company share an office for editorial and design work; but occasionally both of them have simultaneous meetings or Anita needs to spread out the pages of a book layout, and their living room becomes an adjunct to the office.

Thus, an efficient workstead routine should include regular housecleaning of living and work areas. And for continued domestic tranquility within a workstead family, these duties must be divided fairly—so that the worksteader isn't called upon to prepare a family meal during business hours or the non-workstead spouse isn't expected to clean the office.

Time Management

Because their businesses are usually of the small and low-budget variety, most worksteaders have too many demands on their time. And, of course, their being at home adds to that problem if household or family duties intrude on them as well. But there are techniques for managing time that can be especially valuable for workstead businesspeople.

The Small Business Administration has published a succinct pamphlet, "Techniques of Time Management," that is worthy of study by any working person. A more detailed treatment of time use is found in Alan Lakein's excellent book, *How to Get Control of Your Time and Your Life.* Both of these publications offer aids to motivation and time management that emphasize close examination of work habits and objectives. For example, worksteaders who keep a time log are certain to be surprised at the way many of their working minutes disappear. In analyzing the results of such a survey, a worksteader probably will find that Lakein's "80/20 rule" applies to him; that is, he uses 80 percent of his work time for tasks that produce only 20 percent of the most important work. (A list of common time wasters appears below.)

Twenty Major Time Wasters

External	*Internal*
1. Telephone interruptions	1. Procrastination
2. Meetings	2. Failure to delegate
3. Visitors	3. Unclear objectives
4. Socializing	4. Failure to set priorities
5. Lack of information	5. Crisis management
6. Excessive paperwork	6. Failure to plan
7. Communication breakdown	7. Poor scheduling
8. Lack of policies and procedures	8. Lack of self-discipline
9. Lack of competent personnel	9. Attempting to do too much at once
10. Red tape	10. Lack of relevant skills

From "Techniques of Time Management," by H. Kent Baker, Management Aids for Small Manufacturers No. 239, Small Business Administration, Washington, D.C., 1979.

Mike Young:

Growing Pains

For six years Mike Young shared his one-bedroom apartment with Mailing List Services, a business that maintains computer mailing lists and supplies labels to about eighty clients. He started on a small scale, with one assistant and little equipment, but as his firm grew, Mike added people, machines, and tape storage racks until his living space was virtually obliterated. "I didn't begin the business to have it in the house; it became a matter of what was feasible," Young says. "It seemed very natural to start out with an office in the living room. And when we needed a key punch, we didn't want to have that in the office area, so we put it in the bedroom so we could close the door on the noise. And then we needed another person to help type, and we put that person at the kitchen table."

Over the course of the six years, Mailing List Services grew on its own, without financial backing from banks or personal investors. "My thinking has always been that I would rather prove to myself that the thing could work before I went to anyone to borrow money," Young states. And on strictly business terms, the venture has made it—in large part because the overhead was minimal in the small apartment. But the strain of eight people (Mike and six other full-time workers, plus one half-time employee) in three rooms became too much for everyone, so Mike has arranged to move the business out of his apartment and into a conventional office building.

For Young, the need to reclaim his living quarters became urgent when the firm added a second two-station key punch machine to the bedroom. "When we put the first machine in there," he recalls, "I put the bed up against the wall and bought a sofa bed. There was barely enough room to open that. Then about six months later we added another two-station machine, and now the bed only unfolds halfway. So I sleep sideways, which has been a real joke."

Obviously these crowded conditions cramp Mike's social life and general comfort, and both he and his business have reached the limit of being able to function adequately in the present surroundings. For the sake of his personal morale and the business's health, Mike Young has concluded that he must leave the workstead behind. "One of the very hardest things for me has been noting the point where the business stops and I begin," Young states. "We are intimate, this creation and I, and at some points maybe too intimate. I don't have a personal life—no wife or kids—and in large part it's because I've paid attention to this thing with its own life. It's been satisfying and it's taught me a lot. But now I'm looking forward to having an office where I can go and close the door and smoke cigars, talk on the phone, and think."

To remedy this inefficient use of time, these experts recommend classifying all tasks according to their importance, assigning an "A, B, C" ranking. The key to using this ranking system is to begin with the *A*'s—the items of highest importance and greatest difficulty. In this way, the worksteader can accomplish two major management objectives: beating the 80/20 syndrome by tackling the toughest jobs first, and discovering which low-priority tasks can be delegated to assistants or temporarily ignored.

In a larger sense, too, time has an almost mystical dimension for some worksteaders. Many people begin home-based careers as a rejection of the time-clock regimentation of conventional office or factory jobs. Consequently, for a certain group of worksteaders, time management seems to threaten the freedom they sought in establishing independent careers. Some of these people probably will continue to blur the lines of business and leisure hours—with the result that they are never quite as organized or effective as they might be as managers—and thus preserve the spirit of flexibility and contentment that a workstead represents for them.

Morale

The example of a worksteader who eschews time-management principles in favor of her sense of freedom also illustrates the individual nature of morale. She may end up working evenings and weekends in the process of avoiding a traditional 9-to-5 schedule, but this lack of a rigid time frame may preserve her independent outlook. In contrast, another worksteader might feel demoralized by the uncertain work hours and unpredictable leisure time of such an arrangement. Thus, the factors that keep each of us satisfied with our working life are mixed in personal proportions, although worksteaders in particular cite several vital ingredients for healthy morale.

People

A workstead setting can accentuate the loneliness of solo professions. Architect John Edwards says that "losing contact with people, even for a couple days, is like being underwater. Something begins to happen to your personality." Edwards and a number of other worksteaders rectify this situation by planning work-related or social activities that provide daily contact with others. Both Edwards and attorney George Hellyer note that they also feel the need to "bounce ideas off" others in their professions, because the isolation of a home office can lead to a narrowing of perspective and a limitation of input. Both men meet regularly with colleagues to share ideas and consult about specific projects.

Contact with professional associates and friends clearly helps morale, but worksteaders also must make time to be with their families. Indeed, family morale is at least partially dependent on the

E. B. Cochran:

A Four-Bedroom Office

Three years ago, management consultant E. B. Cochran and his wife, Sally, purchased a split-level home with a lovely view. They chose it specifically because it had four small bedrooms on the ground floor, which Cochran promptly converted to office space for himself and his four assistants. He knocked out a wall to make one large office for himself, added sound-proofing, a fireplace, sliding doors, and a deck. In the remaining rooms, Cochran installed a computer and work space for the programmer and economist he employs, as well as desks, files, and furniture for two secretaries. "I didn't want to waste so much time going to an office, even nearby in town," he says. "This is just perfect."

Cochran's highly successful consulting business requires him to travel fairly often, but his writing and intensive idea work are done in his elegant workstead office. "The most important thing that happens in this place is when I'm sitting in my chair looking at the ceiling," Cochran relates. "So I want to be comfortable here." He is also a man who is eager to begin a project or tackle a client's problem. "You never know who's going to be on the phone when it rings. It could be a major general in the Air Force or the head of the GSA," he notes. But even when the phone is likely to be quiet—in the evenings and early morning hours—E. B. Cochran is usually at his desk: "I'm a night owl, and I work seven days a week. My work is my pleasure."

workstead businessperson's participation, and mere proximity does not substitute for daily activities together.

Change of Scene

Getting away from the workstead is another requisite for most home-based businesspeople. Occasionally, such escapes should be total—a few days or weeks away from the twenty-four-hour work and living place, and preferably away from the work as well. Apart from these lengthier excursions, worksteaders need a change of scene during their regular work schedules for the same reason that a room needs airing—to get the staleness out.

Writer-publisher Malcolm Margolin often takes work with him to a neighborhood coffee house for the new setting and the company: "I think it's perverse to sit in a room all by yourself for more than a couple of hours, though I've done it a lot. There gets to be a point in the day when I go off for a cup of coffee or tea and take a pile of papers along—and it's just to have a little bit of that human stuff all around me."

Leisure

The very nature of a workstead often seems to conspire against leisure. The office is right behind a door, or across a room, or in some other perilously close spot. And as one worksteader describes it, "The work just seems to chase me around, so lots of times I go back to it when I've sworn I was through for the day." An exclusive diet of work is detrimental to morale over time, however, and most worksteaders have made firm rules for themselves to build some leisure time into each day.

"You just have to sit down and say, 'That's it,'" notes Bill Welch, co-owner of Glad Hand Designs. "No matter how important some order seems to be, you have to set a time and then quit, even if the order has to wait until tomorrow." Welch recalls the moment that taught him this lesson: "I was working as a carpenter—living in the house I was remodeling—and I woke up one morning, and there was my Skil saw next to me, by the bed. And I said to myself, 'Something's wrong here.' From then on, I've made sure there's always some time for myself."

Setting

A workstead is the chosen environment of its occupant, and this "home-grown" element usually adds to the worksteader's enjoyment of doing business there. But the setting for a home office can be a negative influence if it becomes too messy or remains perpetually unfinished. The disorganized state of a workstead can be especially uncomfortable for the worksteader's employees or colleagues, who may wish to improve the work setting but are reluctant to criticize their co-worker's home. Even in a one-person workstead, the neatness and decor have at least a subliminal effect on the occupant's morale.

Fitness

As we all know, physical activity has a positive effect on both work performance and morale. Regular exercise is the "change of air" for the body, and it also can provide contact with other people and a simultaneous change of scene. Activity can serve as a work break, too; editor Ellen Kaiser uses her stationary bicycle when her energy lags: "I used to take a shower to wake up in the late afternoon, but now I pedal for ten minutes, get nice and warm, and save a whole lot of water."

What if...

What if we all stayed home? Today that question is not so far-fetched, even though everyone can't or wouldn't want to work at home exclusively. But the increasing pressures of the go-to-work world and the continuing advances in communications are making worksteads attractive and practical for more of us. There are exciting possibilities for change in our families, jobs, neighborhoods, and society if many more people stayed home to work.

A Family

Family members could begin each day with relative tranquility, rather than expending valuable time and energy on hectic preparations for traveling to an outside work place. While space arrangements and family routines would have to be well organized, workstead members would have greater flexibility to schedule their work and leisure activities. Parents could participate more in child-care groups, school functions, and after-school projects and still meet the demands of their work loads. Couples without children or single people also could take part in weekday activities with children by arranging work routines to set aside a day or a few hours for that purpose. There would be time, too, for a dog walk or other exercise at midday.

A Company

Most working people have stayed home on occasion to finish a report, prepare for a meeting, or catch up on recordkeeping. Employers have recognized the assets of such temporary workstead situations. The home comforts and change of scene from a busy office atmosphere can provide the extra support needed to complete a difficult project or one that requires intense concentration.

Companies with flexible work-at-home policies also are able to retain valuable employees who might otherwise be unable to travel to an office at a certain time in their lives. An excellent example of such a transitional workstead is offered by San Francisco radio station KMEL. One of the FM station's popular disc jockeys, Mary Holloway, was permitted to broadcast her evening program from home for the first six weeks after giving birth to her daughter, Sarah. This arrangement, which was suggested by a KMEL staff member, resulted in good publicity for the station and a month-and-a-half for mother and daughter to be together constantly.

"It was a great experience," Mary recalls. "It was a special time for my husband and me, and the baby was our first priority; so I really appreciated being able to do my show at home." A sophisticated telephone connection linked Holloway with a specially hired engineer at the station; Mary told listeners that she was speaking from her dining room, and Sarah provided occasional sound effects. Now that her baby is older, Mary is broadcasting from the station again, while her husband, musician David Fredericks, takes care of Sarah during Holloway's week-night program.

The future holds an even wider range of alternatives for businesses. In many companies, 50 to 75 percent of the employees could do their work at home. For example, typists, stenographers, computer operators, accounting and payroll clerks, and many managers could work easily from a base other than a central office building; the tools and equipment for their jobs could be installed in residences, as well as communication systems to maintain contact with office staff. To preserve personal contact among workers, a company could schedule weekly meetings at a central gathering place (an office building or other location made available to the firm). The weekly meeting would be a time for individual conferences, team evaluations and planning sessions, and general catching up with fellow employees. Although some key personnel would work in a central headquarters most days, fewer people would be crowding the roads, parking garages, and business space, resulting in significant savings in time, resources, and expense for office facilities. The liberated space in cities and other business centers could be converted to housing and shopping space, adding both life and variety to these places.

A Neighborhood

The pace of neighborhood life would change perceptibly if most of its residents stayed home to work. The hour or so formerly given to a commute could become the time for sweeping sidewalks, watering shrubs, going out for the paper, and meeting the people next door. What used to be only a friendly wave as two station wagons passed at 8:00 A.M. could become a conversation and lead to a sharing of resources and like interests. A genuine interest in the common goals of the neighborhood would be likely to develop. Local issues and ideas could be communicated more effectively, and opinions, votes, and energy could be marshaled readily. A vacant lot or nearby park could serve as a community garden, tended by a rotating crew of planters and weeders who are free to put in a few hours because of their flexible work schedules.

The sense of community that has been diminished by the pace and pavements of modern life could be at least partly reclaimed by a workstead population. If several members of a household work at home, they must learn to get along together, and a large number of such households can create a fabric of cooperation in any neighborhood. This spirit of cooperation and community is essential to a truly healthy society, as urban historian Lewis Mumford has ob-

Violet Anderson:

Word Processing at Home

Violet Anderson is one of the original members of the three-year-old home work program established by Chicago's largest bank. A full-time employee of the trust department of Continental Bank, Violet transcribes dictation on a word-processing machine in the back bedroom of her spacious city apartment. The bank supplied the equipment and the necessary furniture and paid for installation of an extra telephone line to connect Violet's home terminal to the bank's central office. The telephone link is used to transmit the recorded dictation to Anderson's home cassette machine and, with the standard computer-input device, to send the transcribed work to the bank's computer for storage and print-out.

At present, Continental Bank's program utilizes an in-house person to send the dictation to home workers and to monitor the printing, which is done only at the bank's offices. Mary McArthur, supervisor of the home work program, notes that even with the use of in-house staff, the program holds great promise. "We've found that we can keep the people at home as productive as the people here," she observes. "And part of what's unique about our program—because people have been typing at home for years—is the electronic transmission back into the bank, rather than having messengers or the employees themselves responsible for getting the text back to the office. It's done electronically, which makes it have tremendous potential for the future."

Part of that future prospect is a new labor market, says Ron Coleman of the bank's communications department. "One of the problems that the bank and other Chicago businesses have been experiencing is getting qualified workers, especially secretarial and clerical people," he states.

"We realize that there are labor pools out there to tap, and this home work program is a way of expanding our labor force."

Violet Anderson used to be part of the commuting labor force, but now would not go back to working in an office. "This is the greatest thing that ever came into being for me," Violet points out. "It's always been traveling to and from work, waiting in the cold or the sweltering heat." Besides saving the time and aggravation of a bus-and-train commute each day, Anderson finds the working conditions preferable at home. "I feel better. It's a much more relaxing atmosphere," she states. "The pressures are not as great when you're working at home as when you're in the midst of the fluorescent lights and people shuffling and running and standing behind you saying, 'I've got to have this right now.' I can be sitting here at home with gobs of work, but the tension isn't there—it's just more relaxed."

Since the program started in 1978, Anderson has not taken any sick days and has only gone into the bank's downtown offices for biannual evaluation meetings. She makes it a point to get outside every day, though, and uses her two former commuting hours for household chores and new activities. "At five o'clock I'm out the door," Violet notes. "I've already got an hour's jump on everybody—for shopping, going places, everything. I started ceramics classes at night, which is something I never had the time or the energy to do before. When I was at the bank, I was just too tired when I'd come home to even think of a class."

One other benefit of Anderson's workstead situation is the chance to meet more of her neighbors. "It's strange," she observes. "I've lived here for years, but now I'll walk down the avenue and people will talk to me, whether they know me or not. After they've seen me a few times, they say, 'Hi, how are you today?' That's something that never happened before."

served. "The stabilities of the family and the neighborhood are the basic source of all higher forms of morality," Mumford writes, "and when they are lacking, the whole edifice of civilization is threatened."

A Society

A growing workstead population could contribute to the stability and health of a contemporary society in several important ways. The economic benefits of worksteads are clear—in reducing or even eliminating oil imports for commuters' vehicles; in freeing costly office space for a variety of uses; and in slowing down our consumption of other natural resources by scaling down the size of businesses and what they require to function.

The increasing percentage of workstead cottage industries and small-scale businesses also helps diversify the economy. A mixture of large and small, concentrated and decentralized enterprises is crucial in a world where abrupt fluctuations—such as a disruption in petroleum imports—can cause immediate chaos. A diversified economy, with a strong component of self-propelled businesses and industries, will suffer far less during such crises than a society that depends greatly on energy-intensive factories, automobiles, and oil-based products.

At present, the laws and regulations for establishing many types of home-based businesses are restrictive and even discouraging. Yet as local governing bodies begin to recognize the social and economic value of workstead businesses, they are changing the legal picture to one that balances individual opportunity and necessary safety measures. As worksteads continue to flourish, a more favorable climate for work-live situations will be increasingly reflected in the laws of every community.

Finally, worksteads have the potential to "rehumanize"—rather than dehumanize—a society. More worksteads would mean more human interaction and a growing reacquaintance with the advantages of individual initiative. The scale of living could become simpler and more personal—as people gain greater responsibility and control over their work as well as their personal lives, as the fruits of individual effort increase in value and respect, and as we develop, collectively, an approach to life that is consistent with the planet's human and natural resources.

Appendices

Update

As *Worksteads* goes to press, we've learned of new developments for some of the people in its pages, and of the untimely death of one worksteader.

Several months after we visited his lovely home and office, business consultant E. B. Cochran died unexpectedly after a brief illness. He was one of the most energetic and brilliant people we have met, and all of us who have contributed to this book hope that his example will inspire those who meet him here. Our own tribute to Mr. Cochran is the respect and appreciation we will always have for his work and his generosity.

Not surprisingly, the political battle over the ownership of San Francisco's Goodman Building continues at a heated pace. As of late September 1980, the city's redevelopment agency planned to serve Goodman Building residents with eviction notices, and the tenants planned to press their pending lawsuit in the dispute over purchase of the property. Elected officials also have entered the controversy: the city's board of supervisors unanimously passed a resolution stating that the tenants should be allowed to remain in the building until all legal matters are clarified, but the mayor refused to sign this resolution. Although the stalwart and determined artists and their supporters are fighting to keep this unique workstead from commercial development, their options seem to be decreasing in the face of powerful political alliances that favor turning an artists community out of its home.

Filmmaker Bob Charlton recently was awarded the first annual Rosalind Russell Filmmaking Grant, given by the Los Angeles International Film Exposition. Charlton was selected for this award for his film *Survival Run,* which documents the participation of a blind runner and his sighted friend in a cross-country race.

Several worksteaders have changed their locations since we visited them, and three people have temporarily—they assure us—separated their offices and homes. Les and Peggy Michaels found that their home-and-office combination was too crowded after the birth of their third child, so they have rented a nearby house and plan to expand the living quarters attached to their medical office. In the meantime, one of their employees lives in the "home" half of the building, so it is still a workstead business.

Crafts catalog publisher Christopher Weills has moved out of his workstead office for the happiest of reasons—he married Sarah Satterlee, the editor of his crafts newsletter, and now shares a home with her near their office.

A Worksteads Network

The worksteads network is growing, as home-based business-people continue to make innovations in their working lives and as the workstead alternative becomes feasible for more of us.

To help worksteaders stay informed and in touch with each other, we're planning a newsletter with information concerning all aspects of the work-live experience.

If you'd like to contribute information, suggestions, or insights concerning home-career opportunities, or be part of the worksteads network, please write to us. We also welcome feedback about this book. Address all correspondence to: Worksteads, P.O. Box 29464, San Francisco, California 94129.

Jeremy Joan Hewes has been a work-steader for most of her career as a writer and editor. She lives and works in San Francisco, where she maintains the branch office of Island Press, of Covelo, California, and keeps in touch with a growing network of workstead people and resources. Her books include *Build Your Own Playground!*, *Good Cops/Bad Cops* (with E. E. Shev, M.D.), and the forthcoming *Redwoods: The World's Largest Trees.*

David Seligman is a professional photographer who lives and works in his warehouse studio in San Francisco. His projects include both creative and commercial photography; *Worksteads* is the first book to be illustrated with his work.

Georgia Oliva is a graphic designer who works in her San Francisco home. Her work includes projects for the advertising business, as well as for book publishers such as Scrimshaw Press, Island Press, and Presidio Press.

Resources

Specialized Publications

Small Business Reporter, Bank of America, Department 3120, P.O. Box 37000, San Francisco, California 94137. Titles include, among others, "Understanding Financial Statements," "Marketing New Product Ideas," "Avoiding Management Pitfalls," "Handcraft Business," "Bookstores," "Mail-Order Enterprises," and "Mobilehome Parks."

Excellent booklets on the fundamentals of starting a business; also a wide variety of publications devoted to specific careers, many of which can be workstead occupations. Write for publications list.

Small Business Administration Publications, P.O. Box 15434, Fort Worth, Texas 76119. Write for Form SBA–115A (list of free booklets) and Form 115B (low-cost guides).

Useful guides and bibliographies on all aspects of business, with special emphasis on small and moderate-sized enterprises.

Federal Information Centers, in major cities, usually located in main federal office building.

The U.S. Government publishes information on a wide range of subjects, and federal bookstores in large cities feature selections from these publications. The reference and documents room of a major library is a good place to find catalogs or listings of publications from all departments of the U.S. Government, since federal stores have a limited selection of titles and little useful cataloging.

Federal Consumer Information Catalog, available free from the Consumer Information Center, Pueblo, Colorado 81009.

Quarterly catalog primarily of interest to consumers, though useful in evaluating products for use in business and surveying products available in a given market.

Internal Revenue Service publications; tax guides and pamphlets available in IRS offices in many cities, or from the Superintendent of Documents, U.S. Government Printing Office, Washington, D.C. 20403.

See list of useful tax booklets on page 136.

Dun and Bradstreet, 99 Church Street, New York, New York 10007.

Publisher of business advisory booklets and financial reports, and supplier of business services including credit reporting.

Periodicals

Briarpatch Review, 330 Ellis Street, San Francisco, California 94102; one year (four issues), $5.00.

Home-grown journal of small-business owners, many of whom are worksteaders, who emphasize "right livelihood and simple living."

In Business, Box 323, Emmaus, Pennsylvania 18049; one year (six issues), $14.00.

Useful new journal for small-business owners, with emphasis on home-based occupations.

Mother Earth News, P.O. Box 70, Hendersonville, North Carolina 28739; one year (six issues), $15.00.
Emphasis on back-to-the-land occupations, with many ideas for home careers.

Venture, 35 West 45th Street, New York, New York 10036; one year (twelve issues), $15.00.
News for business start-ups, with emphasis on venture capital and new vocations.

Organizations

Small Business Administration, offices in major cities; toll-free phone number: (800) 433-7212, except in Texas, where it's (800) 729-9801.
In addition to its loan programs, the SBA can refer worksteaders to other funding sources, to classes and seminars on business, and to consultants from the Service Corps of Retired Executives (SCORE).

Office of Minority Business Enterprises, offices in major cities and in the Department of Commerce, Washington, D.C. 20230.
Assists minority people in developing businesses, with counseling and referrals to possible funding sources.

National Association of Accountants, 919 Third Avenue, New York, New York 10022, and offices in major cities.
Local chapters of this organization give free advice to small-business owners concerning recordkeeping and related procedures.

Artists Equity Association, 3726 Albemarle Street, N.W., Washington, D.C. 20016, and 81 Leavenworth Street, San Francisco, California 94102.
This organization provides specialized resources and publications for artists, with much material on the legal aspects of work-live spaces and information helpful to artists in their business dealings. The San Francisco office is assembling a report from its 1979 conference on work-live spaces for artists; write for information.

Associated Council of the Arts, 570 Seventh Avenue, New York, New York 10018.
Cooperative organization of state and local arts councils, emphasizing support for arts programs and information useful to artists, including housing and studio spaces.

Office Furnishings and Supplies

Conran's, 145 Huguenot Street, New Rochelle, New York 10801; also several retail stores in the New York area. Catalog, $2.00.
Mail-order supplier of many useful and moderately priced storage components, as well as a full line of home and office furniture.

Viking Office Products, 13515 South Figueroa Street, Los Angeles, California 90061; free catalog available.

Mail-order supplier of office materials and furnishings.

Day-timer, P.O. Box 2368, Allentown, Pennsylvania 18001; free catalog available.

Mail-order supplier of stationery, preprinted office forms, and recordkeeping and appointment books.

Bibliography

General Information

Berry, Wendell. *The Long-Legged House*. New York: Ballantine Books, 1971 (out of print).

———. *The Unsettling of America*. San Francisco: Sierra Club Books, 1977.

Briarpatch Community. *The Briarpatch Book*. San Francisco: New Glide/Reed Books, 1978.

Bureau of the Census. "Selected Characteristics of Travel to Work in Twenty-one Metropolitan Areas: 1976." Washington, D.C.: U.S. Government Printing Office, 1979.

de Moll, Lane, and Coe, Gigi. *Stepping Stones: Appropriate Technology and Beyond*. New York: Schocken Books, 1978. (See also *Rainbook: Resources for Appropriate Technology*, a companion volume by the editors of *Rain* magazine, Schocken Books, 1977.)

Farallones Institute. *The Integral Urban House: Self-Reliant Living in the City*. San Francisco: Sierra Club Books, 1979.

Gilliam, Harold. "A Totally Different Era Is Coming." San Francisco *Chronicle*, February 3, 1980, p. 37.

Howard, Jane. *Families*. New York: Simon & Schuster, 1978.

Jacobs, Jane. *The Death and Life of Great American Cities*. New York: Random House, 1961.

———. *The Economy of Cities*. New York: Random House, 1969.

Johnson, Warren. *Muddling Toward Frugality*. San Francisco: Sierra Club Books, 1978.

Mumford, Lewis. *The Culture of Cities*. New York: Harcourt, Brace and Company, 1938.

———. *The Urban Prospect*. New York: Harcourt, Brace & World, 1968.

Needleman, Jacob, editor. *Speaking of My Life*. New York: Harper & Row, 1979.

Roszak, Theodore. *Person/Planet*. Garden City, New York: Anchor Press/Doubleday, 1978.

Schumacher, E. F. *Small Is Beautiful*. New York: Harper & Row, 1973.

Swatek, Paul. *The User's Guide to Protection of the Environment*. New York: Ballantine Books, 1970.

Terkel, Studs. *Working*. New York: Pantheon Books, 1972.

Todd, John, and Todd, Nancy Jack. *Tomorrow Is Our Permanent Address*. New York: Harper & Row, 1980.

Toffler, Alvin. *The Third Wave*. New York: Morrow, 1980.

Business Operations

Baker, H. Kent. "Techniques of Time Management." Small Business Administration, Management Aids for Small Manufacturers No. 239, Washington, D.C., 1979.

Beatty, Jim; Kamoroff, Bernard; and Honigsberg, Peter. *We Own It; Starting and Managing Co-ops, Collectives and Employee-Owned Ventures.* Laytonville, California: Bell Springs Publishing, 1980.

Cotton, John. "Keeping Records in Small Business." Small Business Administration, Small Marketers Aid No. 155, Washington, D.C., 1974.

Dible, Donald M. *Up Your Own Organization!* New York: Hawthorne Books, 1974.

Kamoroff, Bernard. *Small-Time Operator.* Rev. ed. Laytonville, California: Bell Springs Publishing, 1980.

Lakein, Alan. *How to Get Control of Your Time and Your Life.* New York: Wyden, 1973.

Levinson, Jay Conrad. *Earning Money Without a Job.* New York: Holt, Rinehart and Winston, 1979.

Michaels, Richard. *Money-Making Business Opportunities.* Englewood Cliffs, New Jersey: Parker Publishing, 1978.

Radics, Stephen P., Jr. "Steps in Meeting Your Tax Obligations." Small Business Administration, Small Marketers Aid No. 142, Washington, D.C., 1975.

Simon, Julian L. *How to Start and Operate a Mail-Order Business.* 2d ed. New York: McGraw-Hill, 1976.

Winston, Stephanie. *Getting Organized.* New York: Warner Books, 1979.

Winter, Meridee Allen. *Mind Your Own Business, Be Your Own Boss.* Englewood Cliffs, New Jersey: Prentice-Hall, 1980.

Home Offices and Business

Blumenthal, Deborah. "At Work They Are at Home." New York *Times,* July 15, 1979, Real Estate Section, p. 1ff.

Davis, Melinda. *Storage: A House and Garden Book.* New York: Pantheon Books, 1978.

Henry, Leon, Jr. *The Home Office Guide.* Scarsdale, New York: Home Office Press, 1968 (out of print).

Ingle, Kay Dockins. "Shopsteading Provides a Home for Entrepreneurs." *Venture,* July 1980, pp. 32–35.

Jacopetti, Roland; VanMeter, Ben; and McCall, Wayne. *Rescued Buildings.* Santa Barbara, California: Capra Press, 1977.

"Movin' On Up." *Venture,* May 1980, p. 7.

Stratton, Jim. *Pioneering in the Urban Wilderness.* New York: Urizen Books, 1977.

Sunset magazine, editors of. *Sunset Ideas for Storage.* Menlo Park, California: Lane Publishing, 1975.

Technology and Communications

"And Man Created the Chip." *Newsweek,* June 30, 1980, pp. 50–56.

Martin, James. *The Wired Society.* Englewood Cliffs, New Jersey: Prentice-Hall, 1978.

Moore, Marilyn. "On the Trail of Low-Cost Long Distance Phone Services." *In Business,* March–April 1980, pp. 21–23.

Thomas, Susan A. "Early Entry in Personal Computer Market." San Francisco *Business Journal,* January 7, 1980, pp. 12–13.

Weiner, Solomon. *Mastering Business Letter Writing.* New York: Simon & Schuster, 1978.

Zoning, Building Codes, and Licenses

Crawford, Ted. *The Legal Guide for the Visual Artist.* New York: Foundation for the Community of Artists (280 Broadway, Suite 412, New York, New York 10007; send $10.50), 1978.

Granville, S.; Laing, K.; and Morris, J. "Zoning and Joint Live-Work Space: A History, and Problems of the Contemporary Artist." Research report, on file at Artists Equity, 81 Leavenworth Street, San Francisco, California 94102.

Lehmann, Phyllis, and Fain, Kenneth. "Staying Aloft." *Cultural Post,* May–June 1979, pp. 1 and 8–9.

Permit and License Handbook. City of Oakland, California, Business Development Division, City Hall, 14th and Washington Streets, Oakland, California 94612.

Project editor: Linda Gunnarson
Consulting editor: Andrew Fluegelman
Designer: Georgia Oliva
Additional photographs: David Burton (pages 91 *right* and 135) and
    Andrew Fluegelman (pages 56–59, 117, and 160 *top* and *lower right*)
Photographic prints: Emilio Mercado
Drawings on pages 104–106: John Edwards (originals), Phillip McDonel
    (finals)
Researcher: Sally-Jean Shepard
Proofreader: Frances Haselsteiner
Typesetting and mechanical production: Community Type & Design of
    Fairfax, California. Set in ITC Garamond. Production supervision
    by Janis Gloystein.
Display type: Omnicomp of San Francisco, California
Printing and binding: Kingsport Press of Kingsport, Tennessee